The Fourth Stat

An Introduction to l

Second Edition

The Fourth State of Matter

An Introduction to Plasma Science

Second Edition

Shalom Eliezer

Plasma Physics Department
Soreq Nuclear Research Center
Yavne, Israel

and

Yaffa Eliezer

Weizmann Institute of Science
Rehovot, Israel

Institute of Physics Publishing
Bristol and Philadelphia

British Library Cataloguing-in-Publication Data

A catalogue record of this book is available from the British Library.

ISBN 0 7503 0740 4

Library of Congress Cataloging-in-Publication Data are available

First edition 1989

Commissioning Editor: John Navas
Production Editor: Simon Laurenson
Production Control: Sarah Plenty
Cover Design: Frédérique Swist
Marketing Executive: Colin Fenton

Published by Institute of Physics Publishing, wholly owned by The Institute of Physics, London

Institute of Physics Publishing, Dirac House, Temple Back, Bristol BS1 6BE, UK

US Office: Institute of Physics Publishing, The Public Ledger Building, Suite 1035, 150 South Independence Mall West, Philadelphia, PA 19106, USA

Typeset by Academic + Technical, Bristol
Printed in the UK by J W Arrowsmith Ltd, Bristol

Dedication

To our four children, Yosi, Lori, Orit and Dalya,
their spouses and all our grandchildren.

Where there is no vision, the people perish

Book of Proverbs, Chapter 29, 18.

Contents

Foreword to the Second Edition

To invade The Fourth State of Matter and to present it in a popular comprehensive book is not an easy task. This might explain why very few, if any, popular books on plasma science are available on the market today. Books of this type are not only interesting but also very useful to the general public as well as to students, engineers and scientists. To our satisfaction the first edition is still being cited on the Internet by leading prominent web sources such as *Encyclopedia Britannica* and NASA. After the first edition was sold out we received many requests for copies of the book. The above encouraged us to write the second edition of *The Fourth State of Matter.*

The aim of the second edition is to bring *The Fourth State of Matter* up to date in the light of progress. Sometimes progress is almost insignificant in ten years and sometimes, as in the field of plasmas, it is immense. The second edition includes a multitude of new discoveries and applications. A new chapter on 'plasma in industry' is added and all other chapters are updated and enlarged. A history of plasma is described throughout the book and summarized in a separate chapter. A bibliography has also been added. This edition contains almost all aspects of plasma science, plasma in space, industrial and energy applications.

The second edition is a nonmathematical book that can be read with negligible previous knowledge of physics. It is not necessary to 'speak physics' to understand and enjoy it.

Yaffa and Shalom Eliezer
Rehovot
January 2001

Acknowledgments

We would like to thank Dr Yehuda Paiss for an exchange of valuable and exciting discussions. His critical reading of the new chapter on plasma in industry is also greatly appreciated.

Special thanks to Edo Dekel for his good advice and help with the technical problems in preparing the manuscript.

Prologue

When my daughter, Lori, began to study physics in high school, she very soon became frustrated and confused with the subject. My husband, who is a physicist and co-author of this book, spent many hours helping her with her studies and tried to impress upon her the importance and necessity of learning this fundamental subject. He tried patiently to explain the complicated formulas in a simplified manner. At the same time he included some pictorial and easy-to-remember comparisons with events of everyday life and some background history and 'gossip' in order to make the subject more captivating and comprehensible. I, myself, who had never studied physics, sympathized with her and could well understand her frustration and irritation as I watched them work out some lengthy and complicated problems on paper. Still, I found myself eavesdropping on his simple comparisons and amusing 'gossip'.

My first encounter with baffling terminology and complicated and lengthy equations was when I was hired as an English typist at a research center. Later, when I became the secretary to the Plasma Physics Department, my husband, who was at that time the head of the department and my boss, spent many hours explaining some of the experiments and basic principles of physics to me. I was also fortunate to work with some very interesting and clever scientists who patiently explained their complicated research to me. Although they tried to stress to me the beauty, romance, excitement and importance of their work, I'm afraid that they failed to excite my curiosity and most of the time I felt excluded from their enthusiasm and involvement.

When, a few years later, I married 'my boss', the head of the department, he encouraged me to attend some popular physics lectures and to read some 'easy' material on the subject. We would later spend many evenings discussing the various topics. The more he explained, the more I pressed him for more, always insisting that he use 'simple English'. I must admit that at times I monopolized his time and exhausted his stamina. But slowly I became more familiar with some of the terminology and found myself becoming involved in some discussions in which I would not have dared indulge in the past. I was often flattered when I met some of my husband's colleagues and, after an hour of discussion, they asked whether I, too, was a physicist.

As secretary to the Plasma Physics Department, I was very surprised at the response I received when I answered the telephone and gave the name of the department. Most of these callers were unaware that there exists a plasma in physics, though they had some basic knowledge of the plasma in blood. 'What does blood have to do with physics?' I was often asked.

I sometimes wonder how the word secretary first originated. I presume it comes from the word secret, as some dictionaries define the word secretary as 'confidential clerk'. As the secretary to a scientific department of over 30 workers, mainly scientists, I was the center for complaints, confidences, advice and so on. Thus many of my co-workers would come to cry on my shoulder. At times our center would arrange visits from prominent investors for certain research projects. These fund providers and senior official clerks seldom had a proper physics background and therefore did not speak 'physics'. It was thus very difficult for the scientists who speak 'physics' to explain to these fund providers, who speak 'English', the importance of a certain piece of research which they feel is essential. I often heard complaints of frustration from my fellow co-workers who had their brilliant proposals rejected because the funder didn't understand the importance of or necessity for such projects.

Whilst on a home visit to Montreal, Canada, I spent some evenings with my childhood friends. Their knowledge of physics was even less than mine. When I told them the name of the department in which I worked, they raised their eyebrows at 'plasma physics'. I gave them my simple explanation of 'plasma in physics' in the following way: 'Plasma in science is a gas. We know that there are three states of matter. This we learn in public school. These are solid, liquid and gas. But there is a fourth state, which is also in the form of a gas. This fourth state is called plasma. When you heat a solid (such as a cube of ice), it turns into a liquid (water). If the liquid is heated some more, it turns into a gas (steam), and by further heating up the gas, you get a different kind of gas (plasma).' My friends were pleased with my very simple and primitive explanation and told me that they had finally learned something. I felt very proud of my ability to enlighten them, if only slightly, on this complex topic; but when I related my simple explanation to my husband and brother-in-law (who is an engineer and well read in physics), they both laughed. Today, my husband uses my simple introductory explanation whenever he lectures to people who don't speak 'physics'.

The following week my sister hosted a small celebration in honor of my homecoming and invited my friends. My husband decided to improve and elaborate on my previous explanation on plasma. He sought out my friends and began a 'physics' explanation of the ionization process involved in plasma. Before he was half-way through, my friends cried off and told him that they preferred my explanation. 'You see', they told him, 'we don't speak "physics".'

I feel that it is important to stress the fact that physicists speak 'physics'. It is very hard for them to explain to the ordinary housewife or to a passer-by some of the topics in physics, without going into their complicated terminology. Without their mathematical equations, without their sophisticated graphs, without their formulas, without their big and small numbers, they are lost for words. This is why the gap between the important administrator and fund provider and the scientist is so vast. This lack of communication not only causes frustration, but sometimes prevents discoveries or the development of very important research.

Following our visit to Montreal, my husband and I and our four children spent a sabbatical year in Austin, Texas. I had the opportunity to get together with many physicists' wives. When I asked them how they coped with questions relating to their husbands' work, most of them said that they would simply reply that they didn't speak 'physics'.

During our sabbatical, I spent time reading popular books on physics. I kept asking myself how this subject could be made more comprehensible to the average individual. I found that while reading some topics, I could easily write about them in rhyming verse. When I re-read those verses, I was amazed to learn that even I was able to understand these topics better. After I had compiled several poems (which were carefully 'censored' by my husband), my husband showed some of them to a few scientists who thought they were 'very cute and charming'; to some passers-by who thought they were 'very informative'; and to some intellectuals who thought they were 'very good, indeed'.

I remember one particular occasion while on sabbatical. We were attending a reception. My husband was approached by a journalist who asked him to explain his field of work—the nature of plasma and how it is used in nuclear fusion, which is such a big issue with the public at large. At the end of a scientific explanation with some relevant numbers and exact formulations, I noticed that somehow my husband hadn't been able to get through to him. I offered my simple English explanation and suddenly his face lit up and he exclaimed that he finally began to understand my husband's explanation.

Now, what has this long prologue to do with the writing of this book? My husband and I decided to write a simple book in 'English' and not in 'hard physics' to 'invade the inscrutable' and to introduce the plasma in physics to the ordinary individual, to the scientist's wife, her friends, some high school students and perhaps even to the funds provider, who all, like myself, do not speak 'physics'. As the world progresses, some solution to the desperate energy crisis must be found. Scientists today believe that nuclear fusion could be the best solution. It is thus not only degrading, but also dangerous that plasma physics remains unknown to the public at large. In our opinion, it is important that this subject be taught more in universities and introduced to the high schools.

This book is the collaboration between a physicist who speaks 'physics' and a secretary who understands 'English'. The physicist explains and the secretary writes, after 'censorship' of the mathematical formulas, sophisticated graphs and incomprehensible numbers. The end result should be understandable to anyone whose knowledge of physics is negligible. This book is not intended for the physicist.

We chose those subjects in physics which are the fundamental ones necessary to the goal of this book—to produce an understanding of plasma in physics and its application for the benefit of mankind. In the following chapters, we hope that, together with us, you will understand some of the basics in physics, topics which you have usually chosen to ignore in the past. Some rhyming verses appear in the appendix, hopefully to enable a better understanding of some of the complicated terminology and phenomena. The purpose of these rhyming verses is to put big ideas and complicated issues into a compact, simplified and sometimes easy-to-remember form. The rhyming verses are by no means intended as poetry, nor do they follow any specific parameters, patterns or metrical forms.

We have omitted the complicated equations, the incomprehensible big and small numbers and the sophisticated graphs. Instead, we have inserted some simple graphs and pictures. We have tried to include some comparisons with everyday life which we hope will facilitate in translating the hard physics into simple English. We believe that these simple, imaginative and picturesque examples will help to make the reading relaxing and will at the same time not only be informative, but will provide a good atmosphere for 'invading the fourth state of matter'.

Yaffa Eliezer
Rehovot, Israel
March 1988

Chapter 1

Highlights to Plasma

1.1 Unveiling Matter

Over 15 billion years ago our Universe was squeezed into an extremely small ball, that was unstable and exploded violently. This was the most gigantic explosion of all time. This description of the early Universe is known today as the 'Big Bang' model. (The Big Bang model is described in more depth in Chapter 3.)

The matter which composed the Universe was so hot that everything was in the form of plasma. Thus, in the very beginning, plasma was the *first* state of matter. The fragments of this explosion became the stars of our Universe, including our own Sun. During the expansion of our Universe, the matter cooled down and thus some of the plasma changed into gas, which further cooled down and became transformed into the liquid and eventually the solid states. This is the reverse of the sequence of events which will be discussed in Chapter 2 on generating plasma as the *fourth* state of matter.

At the beginning of civilization, man was familiar with earth and rocks, water and rain. Naturally, therefore, he identified the solid and the liquid states of matter. Thus we refer today to the solid and the liquid as the first and second phases of matter. A few centuries ago scientists realized that a third state of matter existed; this state is gas. The first physical law for gases was discovered by the English physicist Robert Boyle slightly over 300 years ago. The existence of a so-called fourth state of matter—plasma—was realized only about a century ago.

We can't read without first learning the alphabet; we can't do mathematics without learning its principles and equations; it is difficult to play music without learning scales; and we can't understand plasma without learning some of the fundamental 'physical terms' and established facts. We will, therefore, begin with matter, which, in this book, is the alphabet which will introduce us to science.

We ask ourselves, what is matter? The dictionary says, 'whatever occupies space—that which is perceptible by the senses—a substance'. Matter is the Earth, the seas, the wind, the Sun, the stars, the ground we walk on, the homes we live in, the clothes we wear, the food we eat; everything on Earth, including man himself, is matter.

The unveiling of science began through matter. Millions of years ago prehistoric man, out in the wilderness, coping with the wildlife and struggling for survival, was getting introduced to the beginning of science—matter. He was learning the alphabet of science. He wasn't *interested* in exploring or learning anything about science, but his inner instinct for survival led him then to learn the different ways to use matter for his simple everyday life; this was vital for his survival. He was able to build a fire by rubbing sticks together and this heat kept him warm. He learned to choose between edible and poisonous plants which kept him alive. He made crude tools out of stones for his daily chores and self-defense. Later he discovered different kinds of metal such as tin and copper. He noticed that melting and mixing tin and copper produced bronze. He came across gold that was washed down with the sands and iron from the meteorite fragments that dropped down from outer space. Still later he noticed other materials such as minerals. He was able to improve his caves with the colored minerals found from plants, from the blood of insects and from the glazy semi-transparent substance found in the residue of volcanic eruptions. We can compare prehistoric man using matter for his existence to the child today beginning to learn the techniques of reading and writing without realizing the importance of this learning, and how it will lead to his individual development and to the benefit of society.

The Babylonians used crude beer for sacrificial purposes and the early Egyptians prepared wine. Although they could not explain the fermentation process in producing beer and wine, they did notice that some kind of transformation was taking place. The Phoenicians and other nations learned how to make glass out of sand and how to melt sodium minerals onto glass. With this they made picturesque beads and jars by putting glazes on pieces of stone or quartz. They learned how to dye things from the fact that certain insects and berries stained their fingers.

For thousands and thousands of years, matter was used because it was available. A significant development in the understanding of matter began with the Greeks about 2500 years ago. Hungry for knowledge and burning with curiosity they visited all the far centers of culture to learn about the practical chemistry that was then applied. Through persistence and debate, they proceeded to establish different theories of matter. 'What is matter?' they queried. 'How can it be used in a better way and where does it come from? What is it composed of? What is the Universe made of? Why does man exist?'

From prehistoric man, on through many civilizations, up to the present day, we have come a very long way in the search for matter. Early man merely sought ways to use matter and was content that it was available; his civilized successors seek ways to *understand* it.

The research into matter led to the discoveries of new materials which were incorporated into daily use. The study of matter has taught man how to grow his food, to clothe himself, make tools, clear the wilderness, till the land, light up his homes, build cities, explore different places by sea and air, improve his health, and even soar into space.

Man learned that through the use of matter he was able to produce energy for the purpose of heating, construction, transportation, communication, etc. His living conditions have vastly improved and his standard of living has become very high. But all his comforts and easy living could be shattered if the world's available energy is exhausted. The fact that the gigantic population of today can be fed at all is highly dependent on energy supply. We can obtain energy from matter sources such as oil, coal, gas, etc. However, this supply of raw material is limited. Scientists today are searching for ways of providing new sources of energy so that our civilization can continue to survive. The scientists of today believe that there is a way of producing energy for future generations. As we read on we will learn that future methods for achieving an unlimited source of energy are closely related to the subject of this book.

1.2 Unveiling the Atom

The Greek philosophers, while arguing about the structure of matter, asked what would happen, for instance, if you take matter and split it into smaller pieces? What happens if you take a piece of copper and divide it in half, and then the half into quarters, and then the quarters into eighths and so on? Could this material be divided indefinitely, or would it eventually become such a small bit that it could not be split any further? As the Greeks lacked the proper instruments and laboratories to test their theories experimentally, their logic was based on suppositions or hypotheses only. As scientific logic is based on experimental facts and their reasoning and logic could not be proved experimentally, the Greek theories remained mere arguments. It was very difficult to *prove* whose logic was easier to accept, and so the arguments flew back and forth.

About 430 BC, the Greek philosopher Democritus of Abdera introduced his theory about the existence of the atom where he suggested that matter was made up of tiny particles that were themselves indivisible. He called these ultimate particles *atomos,* the Greek word meaning indivisible. Democritus believed that atoms were in constant motion, that they combined with others in various ways, and they differed from each other only in shape and in arrangement. Today we know that this atomic theory was a good guess. It is unfortunate that he lived 24 centuries before experimental science could prove the concepts of his theory.

With modern experimentation it has been established that a piece of copper, for example, can be divided into atoms. These atoms are the smallest units maintaining the chemical properties of copper. It is possible to divide these atoms of copper further, but they then lose their chemical identity and transform into particles with properties completely different from those of copper. Thus Democritus' theory that the atom is indivisible is not correct; however, there is a smallest indivisible piece which maintains its identity (that is, retains the chemical properties of copper).

The well known philosopher Aristotle bitterly attacked Democritus' theory. Aristotle's philosophy of the material world was based on primitive matter consisting of only four elements: water, air, fire and earth. Each of these elements possessed two properties out of the following four media: hot, cold, wet and dry. Aristotle believed that cold and dry were combined to form earth; cold and wet to form water; wet and hot to form air; hot and dry to form fire. In this theory it is possible to go from one element to another through the medium of the properties they possess in common (see figure 1.1). From Aristotle's misleading theory, the alchemists formed their own understanding of the existence of matter. They introduced the transmuting agent called the Philosophers' Stone, which, if produced, could turn base metals into gold and also become man's perfect medicine, *the elixir vitae,* or elixir of life. Although today we can laugh at alchemy as a mere fool's search, its fundamental

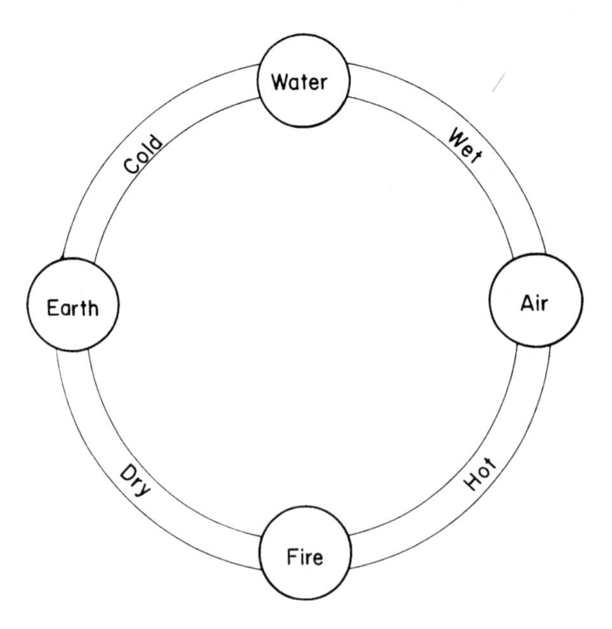

Figure 1.1 Aristotle's material world.

principle—that all kinds of matter had a common origin and could be transmuted from one to another—bears a resemblance to the concept of unity of matter held in physics today. Science is still grateful to the practice of alchemy. In an effort to prove their beliefs and search for gold, they examined and tested every substance known to man and thus laid down a good deal of basic knowledge of the properties of various chemicals and compounds. Francis Bacon, the brilliant 16th-century Englishman who pioneered the scientific method, gave one of the best descriptions of alchemy's contribution to science: 'Alchemy may be compared to the man who told his sons that he had left them gold buried somewhere in his vineyard; where they by digging found not gold, but by turning up the mould about the roots of the vines, procured a plentiful vintage. So the search and endeavours to make gold brought many useful inventions and instructive experiments to light.'

Thus, from the time of Democritus, the idea of atoms was pushed aside for some 2000 years and ignored. Then, about 300 years ago, some famous scientists began seriously to reconsider the idea of the atoms. The Italian scientist Galileo Galilei revived Democritus' theory in the beginning of the 17th century. Then, in 1803, John Dalton performed many experiments and concluded that all matter was made up of indivisible atoms. Therefore, if we take a piece of any element such as copper, for example, and split it into smaller and smaller pieces, the smallest piece that we would finally obtain—one that we could not split any further (and still retain the material copper)—would be the *atom* of copper. Atoms of copper would be different from atoms of gold or tin or other materials. It was thus concluded that every element contains its own kind of atoms which are the same in every sample of that element but are different from the atoms of different elements. With Dalton's great contribution of assigning a specific weight to the atom of each element, the natural elements were each given a unique atomic weight.

Chemists have thoroughly investigated the different properties of solids, liquids and gases and found that the smallest constituent of any material that retains its chemical properties is usually a molecule or an atom. A molecule is composed of two or more atoms. For example, air, which is a gas, is mainly composed of molecules of nitrogen and oxygen. Each molecule of nitrogen contains two atoms of nitrogen and each molecule of oxygen contains two atoms of oxygen. The smallest constituent of water is a molecule containing two atoms of hydrogen and one atom of oxygen. In general, each molecule contains integer numbers of each specific atom. However, some metals, such as copper, iron, gold, tin, silver, etc, are composed of atoms and not molecules. It is extremely hard to imagine the small size of a molecule or an atom. Scientists have found that in a grain of sand or any other material of similar size, there are billions of billions of atoms. More specifically, in a one centimeter length, one could arrange about one hundred million atoms in a row.

It would be too tedious and lengthy to go into all the great achievements and findings of matter leading to the understanding of the atom. However, the most important breakthrough came with the writing down of the Periodic Table in 1869, by Dimitry Ivanovich Mendeleyev, a Russian chemist. In 1871 he published an improved version where he left gaps, forecasting that they would be filled by elements not then known. This table is a unique listing of all the chemical elements in order of increasing weight of the atom.

During Mendeleyev's time not all the elements of today were known. The empty gaps for missing elements that he left in his table have been filled as new elements have come to light. All matter is made up of about 100 elements which are the basic blocks from which we and our surroundings are constructed. For example, our bodies contain long and complicated chains of carbon, hydrogen and oxygen blocks, as well as other compositions. We can look at the matter surrounding us as a puzzle made up of the blocks of elements. For some matter the puzzle has a small number of pieces, while for others the puzzle can be very large. In order to put the puzzle together we have to put the different pieces into place. Mendeleyev defined and arranged the pieces of the puzzle for the different elements in such a clever order that the puzzle of matter can easily be put together. It is amazing that modern science has not changed the order that Mendeleyev imposed on the basic blocks of matter. Mendeleyev's table is presented here in modern form as table 1.1.

In this table there are seven horizontal lines and 18 vertical rows which are actually denoted by eight vertical groups (IA, IB, etc). The horizontal lines in Mendeleyev's table represent the cycle while the vertical rows represent the chemical properties. The elements in each vertical group have similar chemical properties. For example, the hydrogen, the lithium, the sodium, etc from the first row are chemically similar, although they are different elements. Correspondingly, in each row the elements behave similarly when interacting with other elements in forming molecules. For example, an element from row number IA can be combined with an element in row VIIA to form a molecule. (Note that $1 + 7 = 8$.) An element of row IIA can be combined with an element in row VIA to form a molecule. (Note: $2 + 6 = 8$.) Moreover, two identical atoms of row IA (or more generally atoms from two elements of row IA) can combine with one element of row VIA to form a molecule, e.g. two hydrogen and one oxygen give the molecule of water. (Note: $1 + 1 + 6 = 8$.) As we can see, many different combinations are possible and elements can be combined from various rows to form molecules. The elements in the last row (0), which are called the noble gases, do not interact with any of the other elements to form molecules. It appears that the magic number in

Table 1.1 Mendeleyev's Periodic Table of the elements.

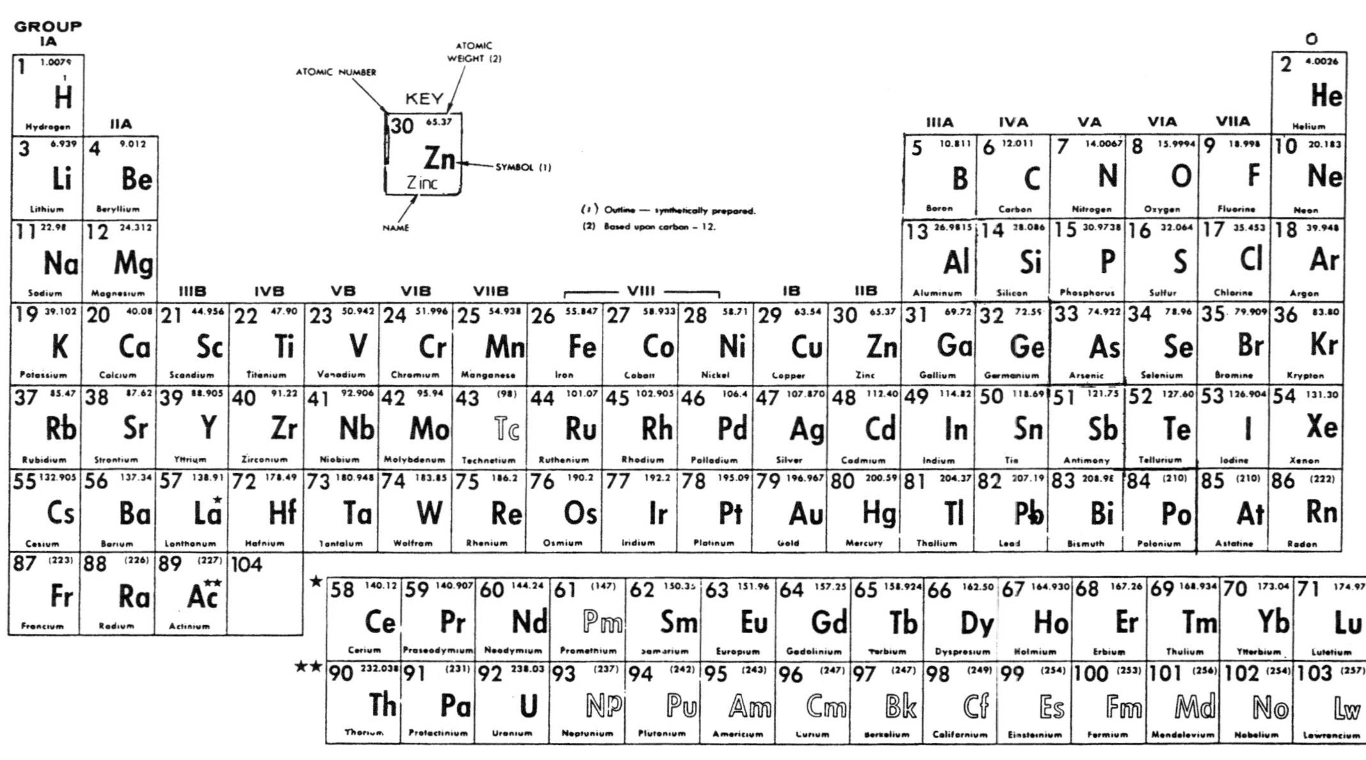

GROUP IA	IIA	IIIB	IVB	VB	VIB	VIIB	VIII	VIII	VIII	IB	IIB	IIIA	IVA	VA	VIA	VIIA	O
1 1.0079 H Hydrogen																	2 4.0026 He Helium
3 6.939 Li Lithium	4 9.012 Be Beryllium											5 10.811 B Boron	6 12.011 C Carbon	7 14.0067 N Nitrogen	8 15.9994 O Oxygen	9 18.998 F Fluorine	10 20.183 Ne Neon
11 22.98 Na Sodium	12 24.312 Mg Magnesium											13 26.9815 Al Aluminum	14 28.086 Si Silicon	15 30.9738 P Phosphorus	16 32.064 S Sulfur	17 35.453 Cl Chlorine	18 39.948 Ar Argon
19 39.102 K Potassium	20 40.08 Ca Calcium	21 44.956 Sc Scandium	22 47.90 Ti Titanium	23 50.942 V Vanadium	24 51.996 Cr Chromium	25 54.938 Mn Manganese	26 55.847 Fe Iron	27 58.933 Co Cobalt	28 58.71 Ni Nickel	29 63.54 Cu Copper	30 65.37 Zn Zinc	31 69.72 Ga Gallium	32 72.59 Ge Germanium	33 74.922 As Arsenic	34 78.96 Se Selenium	35 79.909 Br Bromine	36 83.80 Kr Krypton
37 85.47 Rb Rubidium	38 87.62 Sr Strontium	39 88.905 Y Yttrium	40 91.22 Zr Zirconium	41 92.906 Nb Niobium	42 95.94 Mo Molybdenum	43 (98) Tc Technetium	44 101.07 Ru Ruthenium	45 102.905 Rh Rhodium	46 106.4 Pd Palladium	47 107.870 Ag Silver	48 112.40 Cd Cadmium	49 114.82 In Indium	50 118.69 Sn Tin	51 121.75 Sb Antimony	52 127.60 Te Tellurium	53 126.904 I Iodine	54 131.30 Xe Xenon
55 132.905 Cs Cesium	56 137.34 Ba Barium	57 138.91 La* Lanthanum	72 178.49 Hf Hafnium	73 180.948 Ta Tantalum	74 183.85 W Wolfram	75 186.2 Re Rhenium	76 190.2 Os Osmium	77 192.2 Ir Iridium	78 195.09 Pt Platinum	79 196.967 Au Gold	80 200.59 Hg Mercury	81 204.37 Tl Thallium	82 207.19 Pb Lead	83 208.98 Bi Bismuth	84 (210) Po Polonium	85 (210) At Astatine	86 (222) Rn Radon
87 (223) Fr Francium	88 (226) Ra Radium	89 (227) Ac** Actinium	104														

*	58 140.12 Ce Cerium	59 140.907 Pr Praseodymium	60 144.24 Nd Neodymium	61 (147) Pm Promethium	62 150.35 Sm Samarium	63 151.96 Eu Europium	64 157.25 Gd Gadolinium	65 158.924 Tb Terbium	66 162.50 Dy Dysprosium	67 164.930 Ho Holmium	68 167.26 Er Erbium	69 168.934 Tm Thulium	70 173.04 Yb Ytterbium	71 174.97 Lu Lutetium
**	90 232.038 Th Thorium	91 (231) Pa Protactinium	92 238.03 U Uranium	93 (237) Np Neptunium	94 (242) Pu Plutonium	95 (243) Am Americium	96 (247) Cm Curium	97 (247) Bk Berkelium	98 (249) Cf Californium	99 (254) Es Einsteinium	100 (253) Fm Fermium	101 (256) Md Mendelevium	102 (254) No Nobelium	103 (257) Lw Lawrencium

the 'periodic game' is 8 and its multiples (16, 24, 32, etc). For example, two atoms of aluminum (row IIIA) can combine with three atoms of oxygen (row VIA) to form a molecule (note: $2 \times 3 + 3 \times 6 = 24$), etc. The 'eightfold' trend holds true also for the vertical B rows; however, with the elements in these rows there are more irregularities than in the A rows. This structure and the irregularities can be understood in the context of quantum mechanics.

1.3 Unveiling the Electron

As far back as the time of the philosopher Thales (600 BC), the Greeks knew that when amber (a brownish yellow substance that came out of pine trees) was rubbed with a cloth, the amber became capable of attracting small bits of paper to it. This is because when amber is rubbed with the cloth, a certain force is created between its surface and that of the bits of paper. This force is an 'electric' force. Amber in Greek is *elektron* and this is how the electric force and later the electron acquired their names.

What is electricity? Electricity is a quantity of electric (or charged) particles, called electrons, either in motion or at rest. When the electrons move, an electric current exists. When they are attached to one atom, the electricity is said to be static.

In the 18th century scientists began to predict that electricity, like matter, might consist of tiny units. They soon learned that electricity existed in two varieties which were called positive and negative.

A current can flow across a wire or through some solutions (such as sodium chloride in water) or across a gap in a vacuum tube (a sealed device in which most of the air has been removed) connected to a battery or any other source of electricity.

When one connects a light bulb to a battery in a closed circuit, a current flows across the wire inside the light bulb (and visible light is emitted). The electrical current (measured in amperes) is proportional to the potential (measured in volts) of the battery; this is known as Ohm's law, named after the German physicist Georg Simon Ohm who suggested this in 1826. It was later discovered that the electrical current flowing across the wire inside the bulb is made up of electrons only. Moreover, the currents transferred from an electric power plant to individual outlets are also composed of electrons only.

In 1832 the famous English physicist and chemist, Michael Faraday (who is considered to be one of the greatest experimentalists of all time and whose important contributions to electromagnetic induction paved the way to the use of electricity today), developed the laws of electrolysis. These were based on the following experiment. Two separated metal rods which are connected to a battery are inserted into a solution. As we know, the battery possesses two poles (the terminals of the electric cell). The rod which was connected to the positive (+) pole of the battery was called by Faraday the anode and the one connected to the negative (−) pole was called the cathode. If one takes, for instance, a solution of sodium chloride in water, a current will flow, while in a sugar solution the current does not flow. Faraday called the liquids in which electricity could flow *electrolytes.* In this case Faraday suggested that *ions* (which are electrically charged atoms or molecules) move through the solutions. There are two kinds of ion, positive and negative; in the sodium chloride solution, the sodium is a positive ion moving towards the cathode and the chlorine is a negative ion moving towards the anode. Thus from Faraday's experiments it was concluded that electricity can produce ions in matter.

The current across a gap in a vacuum tube is set up by placing two separated wires in a closed tube, from which most of the air has been removed. The two wires are connected to a powerful battery. The wire which is connected to the positive pole of the battery is the anode and the other, which is connected to the negative pole, is the cathode. In numerous experiments performed during the 19th century in such vacuum tubes, it was noted that when the current flowed across, there was a greenish glow about the wire that was attached to the cathode of the battery. The rays which began at the cathode ended at the anode. These rays were called 'cathode rays' and were believed to be the electric current. The particles in these rays were later proved to be the negatively charged electrons as they were moving from the cathode to the anode (from the minus to the plus). In these experiments the scientists found that the current in the vacuum tube flowing across from the cathode towards the anode is not composed of ions, as in Faraday's experiments, but rather of streams of electrons. Furthermore, in the vacuum tube experiments, the current flowing from the anode to the cathode (from the plus to the minus) was found to be composed of positive ions.

From the above three experiments, namely, the electric current through a light bulb, the electric current in solutions and the electric current in vacuum tubes, it was concluded that: (a) the flowing current across a conducting wire is composed of electrons only; (b) the flowing current in solutions is composed entirely of positive and negative ions; and (c) the flowing current in vacuum tubes is made up of electrons and positive ions.

The English physicist, Sir William Crookes, in 1879, while considering the unusual properties of gases in the electrical discharges in closed tubes as described above, suggested that these gases are the 'fourth state of matter'. Furthermore, in 1885, Crookes inserted two tiny rail tracks inside a vacuum tube and placed a small propeller which was capable of moving freely on the tracks. When he switched on the circuit the cathode rays began to stream across the tube and he noticed that the propeller began to turn and move along the track. This seemed to show that the cathode rays possessed mass (therefore, they were capable of applying a force to turn the propeller) and were streams of atom-like particles, rather than a beam of massless light. Moreover, in another experiment, he showed that the cathode rays could be pushed sideways in the presence of a magnet. This meant that, unlike either light or ordinary atoms, the cathode rays carried an electric charge.

Another English physicist, Joseph John Thomson, in 1897 confirmed that the particles making up the cathode rays were charged with negative electricity. The cathode rays were considered to be made up of streams of electrons. Thomson is given credit for having discovered the electron and received the Nobel Prize in 1906 for this discovery.

The German physicist Wilhelm Wien, in 1898, and later J. J. Thomson in 1901 while performing similar experiments with vacuum tubes containing hydrogen gas, identified a positive particle with a mass almost equal to that of the hydrogen atom. The New Zealand-born English physicist Ernest Rutherford showed in 1919 that when the nucleus of nitrogen was bombarded with alpha particles (which will be discussed in Section 1.4) a hydrogen nucleus was obtained. In 1920, Rutherford defined the hydrogen nucleus as a fundamental particle and named it the proton.

After Thomson had proved that all atoms contained an elementary particle called an electron, it was concluded that the atom must also contain particles with positive electric charge to balance the negative charge of the electron. The elementary particles in the atom that carry positive charges were called protons. The number of electrons in the atom must be the same as the number of protons. The total negative electric charge carried by the electrons must balance the total positive charge carried by the protons if the atom is to be electrically neutral. A hydrogen atom is made up of one proton and one electron. Since the electron is very light, the mass of the proton is almost the same as that of the hydrogen atom. Today we know that it would take 1836 electrons to possess the mass of a single proton. Since the electrons are so light, most of the mass of an atom is contained in its core.

The chemical properties of the elements are determined by the electrons in the atom. The electrons in the atom are arranged in shells (spherical layers) in a definite order. Some atoms have more shells than others.

Let's go back to Mendeleyev's table (table 1.1) and look this time at the horizontal lines. The number associated in the table with each element represents the number of electrons in one atom. For example, an atom of copper has 29 electrons while an atom of gold has 79 electrons. The atoms of the first horizontal line, namely hydrogen and helium, have only one shell of electrons. If we look at the element of radium (Ra), which is in the last horizontal line, we will note that it has seven shells.

In the first shell there is room for two electrons only. Since the helium atom already contains two electrons and only one shell (a full shell), it does not have room for interaction with other atoms to form a molecule. Therefore it is a noble gas. In the second shell there is room for eight electrons only. For example, lithium, which has three electrons, has two electrons in the first shell and one in the second shell. Oxygen has eight electrons, where two are in the first shell and six in its second shell. When forming a molecule, the oxygen with its six electrons in the outer shell will attach another two electrons from other atoms (for example two hydrogens) to form a molecule. It is important to realize that the electronic shells have a tendency to become full. Due to the exchange of electrons from one atom to another during the formation of the molecule some atoms acquire positive charge (those which gave up some electrons) and some negative charge (those which received some electrons). The third shell usually has room for 18 electrons. However, if the third shell is the last shell of the atom, it can have only eight electrons and the shell before last cannot have more than 18 electrons. In general any last shell cannot have more than eight electrons. The maximum number of electrons permissible in each shell is shown in table 1.2.

The table shows the maximum number of electrons permitted in each shell. It thus also describes the electron configuration of the noble gases. The first horizontal line describes the electronic structure of the element helium; the second, the element neon; the third, the element argon; the fourth, the element krypton; the fifth, the element xenon; and the sixth, the element radon; the seventh horizontal full configuration is still not completed.

The chemical properties and the formation of molecules are actually determined mainly by the last shell. The atoms in the first vertical row (IA; also IB) have one electron in their outer shell; those in the second row have two electrons in their outer shell, etc, while the noble gases have eight electrons in their outer shell (except for helium, which has

Table 1.2 Maximum number of electrons allowable in shells.

Horizontal line	First shell	Second shell	Third shell	Fourth shell	Fifth shell	Sixth shell	Seventh shell
1	2						
2	2	8					
3	2	8	8				
4	2	8	18	8			
5	2	8	18	18	8		
6	2	8	18	32	18	8	
7	2	8	18	32	32	18	8

only two electrons in total). The 'eightfold' rule described previously can therefore be understood from the fact that atoms combine to form molecules by filling up their last shell, either by acquiring or by giving away electrons.

Let's have another look at table 1.1 and note this time that two rows have been removed and listed separately. As can be seen, the elements containing between 58 and 71 electrons are listed separately. They are referred to as the lanthanide series and are grouped together because they possess similar properties to lanthanum (possessing 57 electrons). This row of elements is denoted by one star. The elements containing between 90 and 103 electrons possess similar properties to actinium (which contains 89 electrons). This series is referred to as the actinide series, and this row of elements is denoted by two stars.

It is also worthwhile to note that not all the atoms are of the same size. Although the dimensions of the atoms are extremely tiny (as mentioned before, one could arrange about one hundred million atoms along one centimeter), they do come in different sizes.

1.4 Unveiling the Nucleus

At the close of the 19th century the German physicist Wilhelm Konrad Roentgen, while working with cathode rays created inside vacuum tubes, found that if he made the cathode rays strike the glass at the end of the tube, a kind of radiance was produced. This radiation was capable of penetrating glass and other matter. It is told that in one of his experiments, he placed his hand accidentally in the way of the radiation. This radiation passed through his hand and struck the photographic plate in his experiment. To his surprise, he saw his bone structure on the film. He named this radiation 'X-rays', the name containing 'X' for unknown. Roentgen was awarded the first Nobel Prize in physics in 1901 for his discovery of X-rays. (Today people go through a similar experience to Roentgen in the above experiment whenever they have any part of their body X-rayed.) Later, physicists found these rays to be light-like radiation made up of waves which have a wavelength much shorter than that of 'ordinary' light.

The French physicist Antoine Henri Becquerel was influenced by Roentgen's discovery of X-rays, and in 1896 took an anode (the rod connected to the positive sign of the battery) of a uranium salt and also performed experiments in vacuum tubes. Afterwards, he placed the anode (uranium salt) in his desk with a photographic plate which was wrapped in black paper (as everyone is aware, film exposed to light becomes ruined). After a few days, he found, to his surprise, that the photographic plate was darkened (ruined). The anode of the uranium had caused the film to become darkened even though it was wrapped up in black paper. He thus concluded that uranium was an endless source of radiation because it gave off continuous rays independently of where it was placed or how it was handled. In 1898, the Polish–French physicist Marie Sklodowska Curie named this radiation 'radioactivity'. Soon thorium was also found to be radioactive. It was also discovered that whatever the radiation was, it was not uniform in properties. For example, in the presence of a magnet some of the radiation curved in a particular direction, while other radiation curved in the opposite direction. Still another part didn't curve at all but moved in a straight line. The conclusion was that uranium and thorium gave off three kinds of radiation. One carried a positive charge, one a negative charge and one no charge at all. Ernest Rutherford called the first two kinds of radiation 'alpha rays' and 'beta rays' after the first two letters of the Greek alphabet. The third was soon called 'gamma rays' after the third letter of the Greek alphabet. The alpha rays and beta rays, which carried electric charges, appeared to be streams of charged particles that are emitted from radioactive elements. In 1900, Becquerel found the beta particles to be identical in mass and charge with the electrons. It was therefore concluded that the beta particles are electrons. These electrons originate in the atom's core and are not the ones moving around the atom.

In 1906, Rutherford learned that the alpha particle carried a positive electric charge that was twice as great as the electron's negative charge. The alpha particle was much more massive than the electron. It was four times as massive as the hydrogen atom and was found to be the nucleus of the helium atom.

Ever since Dalton's revival of the atomic theory, chemists had assumed that atoms were the fundamental units of matter. They believed that they could not possibly be broken up into anything smaller. The discovery of the electron showed that some particles might be far smaller than the atom. The study of alpha, beta and gamma radiation had shown that atoms of uranium, thorium and other elements can radiate smaller particles than the atom. The alpha particle, for example, is about a hundred thousand times smaller than one atom.

For many years it was thought that, like the hydrogen atom, all atoms were made up of only two elementary particles, the proton and the electron. The mass of the nucleus increases faster than its charge. For example, the mass of the nucleus of a hydrogen atom is about 1 (measured in atomic units and referred to as the atomic mass) and its charge is 1 (positive). The mass of helium is 4 while its charge is 2. Lithium has a mass of about 7 while its charge is 3. Since the electron has a tiny mass, it was assumed that the mass of the atom is mainly composed of proton masses. In this picture, the core of the helium atom has 4 protons and 2 electrons and its charge is $4 - 2 = 2$, as it should be. The core of nitrogen, for example, has 14 protons (a mass of about 14 atomic units) and 7 electrons, so that its positive charge is $14 - 7 = 7$. This positive charge is balanced by the negative charge of the electrons moving around its core. This theory ran into many complications because the only thing that would cancel a positive charge, as was known in 1914, was the negative charge—the electron. It was thus supposed that a nucleus would contain about half as many electrons as protons, since they do not affect the mass while they cancel half of the positive charge.

The above proton–electron theory of the nucleus seemed to satisfy the scientists for over 15 years, until the great Austrian physicist Wolfgang Pauli, in 1924, worked out a theory that treated the protons and electrons as though they were spinning on their axes. This spin could be in either direction and he called it the nuclear spin. His theory suggested that the electron and the proton each have a value of spin equal to 1/2. The addition of spin is not the simple arithmetic that we are familiar with. For example, $1/2 + 1/2$ can give either zero or one. By further adding 1/2 spin, one can get either 1/2 or 3/2. In general, adding an even number of spins gives an integer (0, 1, 2, 3,...), while adding an odd number of spins results in half an odd integer (1/2, 3/2, 5/2,...). Therefore, if you have an odd number of particles in the nucleus, you will never find either a zero or a whole number as the spin of the nucleus; this sum will always include a fraction. Consequently, if one measures the spin of a particular atomic nucleus, one can tell at once whether the nucleus contains an even number of particles or an odd number.

The nuclear spin of nitrogen-14 was measured accurately over and over again and it turned out to be 1. There was no doubt that there were an even number of particles in the nitrogen-14 nucleus. Yet the proton–electron theory suggested 14 protons and 7 electrons inside the nucleus, for a total of 21 particles (plus the 7 electrons moving outside the nucleus), implying an odd number of particles inside the nucleus, while the experimental nuclear spin of nitrogen-14 indicated an even number.

With the establishment of the nuclear spin, scientists soon began to realize that one of their theories was wrong. But which one? As the nuclear spin was a matter of accurate measurement which could be repeated over and over again, while the proton–electron idea was a theory only, scientists soon began to realize that the proton–electron theory had to be wrong. This last development led to the search for a third particle—the neutron.

Throughout the 1920s scientists searched for the neutron but without success. One of the problems was that this particle was electrically neutral. Particles could be detected in a variety of ways, but every single way makes use of their electric charge. Charged particle rays can leave a detectable track of atoms or molecules which have acquired or have given up electrons. Such particles (ions) are easy to detect. The suggested neutron however, which was not a charged particle, was not expected to be able to produce ions. It would wander among the atoms without either attracting or repelling electrons and would therefore leave the atomic structure intact. The neutron was invisible and the search for it seemed hopeless.

Then, in 1930, the German physicist Wilhelm Georg Bothe and a co-worker, H. Becker, were bombarding the light metal beryllium with alpha particles. They expected some protons to be knocked out of it, but no protons appeared. However, they did detect some sort of radiation. Bothe and his co-worker tried putting objects in the way of this radiation and found it to be remarkably penetrating. It even passed through several centimeters of lead. The only form of radiation that was known at that time to come out of bombarded matter with the capacity of penetrating a thick layer of lead was gamma rays. They therefore decided that they had produced gamma rays and reported this. In 1932 the French scientists Irene Curie and Frederic Joliot (known as the Joliot-Curies) repeated the same work and got the same results. Among the objects that they placed in the path of the new radiation, they included paraffin, which is made up of the light atoms of carbon and hydrogen. To their surprise, protons were knocked out of the paraffin. Gamma rays had never been observed to do this before, but they didn't know what else the radiation might be and they simply reported that they had discovered that gamma rays were capable of a new kind of reaction.

In that same year, the English physicist James Chadwick claimed that gamma rays, which possessed no mass, simply lacked the momentum to knock a proton out of its place in the atom. Even an electron (with its small mass) was too light to do so. Any radiation capable of knocking a proton out of an atom had to consist of particles that were themselves pretty massive. Chadwick stated that this was the 'missing' neutron particle. Chadwick managed to work out the mass of the neutron from his experiments and by 1934 it was quite clear that the neutron was slightly more massive than the proton. The fact that the neutron was just about as massive as the proton was what would be expected if the neutron was a proton–electron combination. It was also not surprising that the isolated neutron eventually breaks up, giving up an electron and becoming a proton. However, the neutron is really not a proton–electron combination. A neutron has a spin of 1/2 while a proton–electron combination has a spin of either 0 or 1. The neutron, therefore, must be treated as a single uncharged elementary distinctive particle.

As soon as the neutron was discovered, the German physicist Werner Karl Heisenberg revived the notion that the nucleus must be made up of protons and neutrons, rather than protons and electrons. This new proton–neutron theory accounts well for the mass numbers and atomic numbers. It was now possible to define the mass number of a nucleus in modern terms. It is the number of protons plus neutrons in the nucleus, while the atomic number is just the number of protons.

Today's accepted structure of the atom is that the atom is like a small cotton ball composed of protons and neutrons and spinning around this are the electrons. Most of the mass is contained within the nucleus while the outer shell is mostly empty space.

In 1913, Ernest Rutherford bombarded a very thick layer of gold with alpha particles. He found that most of the alphas penetrated the foil of gold without being deflected significantly, that is, they passed through almost in a straight line. A very tiny percentage of the alphas were deflected at large angles. From this experiment Rutherford concluded that the diameter of the nucleus is significantly smaller than that of the atom.

The atom is very, very tiny. With a diameter of less than a hundred-millionth of a centimeter, it's hard to imagine something so unbelievably small. Yet its center, also called the nucleus or the core, is only about a hundred-thousandth of the atom in diameter but contains almost its entire mass. In order to understand these small numbers, let's picture a football stadium. The size of the nucleus is like a grain of wheat in the center of the field while the external electrons moving around are like Olympic runners encircling the circumference of the stadium. The strange fact is that the grain is thousands of times more massive than the athletic runners.

All the atoms of an element have the same number of protons in their nucleus and the same number of electrons in the atom's surrounding cloud, but they may have different numbers of neutrons. Isotope is the name given to two or more atoms with the same atomic number (protons and electrons) but with different atomic weights. In other words, isotopes are members of the same family (element), i.e. they have the same number of protons, but they have a different number of neutrons. For example, oxygen has three isotopes. The atomic number of oxygen is 8. All the isotopes of oxygen have the same number of protons and electrons. Each has eight protons and eight electrons, but the number of neutrons in each isotope varies—one has eight neutrons (the most common), one has nine, and one has ten. Therefore, the mass numbers of the isotopes are 16, 17, 18.

About 70 out of the 100 natural elements on Earth are present in mixtures of two, three, or more isotopes. Radioisotopes have the same chemical properties as those of the regular nonradioisotopes. These are atoms of the same element except that they emit radiation. The chemical properties, as we already know, are determined by the electrons in the atom. The number of electrons is always the same as the number of protons, which is referred to as the atomic number. Alpha, beta and gamma radiation are emitted by the nucleus and the nature of radioactivity is thus related to the atomic weight.

Again, let's return to Mendeleyev's table 1.1. At the upper right-hand corner of each element the atomic weight is listed in atomic units. This system is defined by carbon-12 (which contains six protons and six neutrons) whose weight is 12 atomic units. This is the 'yardstick' for these measurements. The atomic weights, unlike the mass number (the sum of the protons and neutrons in a nucleus), are rarely whole numbers. There are several reasons why the atomic weights are not integers: (a) the neutron is slightly heavier than the proton; (b) the weight of a nucleus is less than the sum of its protons and neutrons (this difference is due to the 'binding energy' which will be discussed in Chapter 3); and (c) the atomic weight is the average weight of the different natural isotopes of an element.

1.5 Unveiling a New State of Matter

To summarize the highlights of this chapter, we can say that the understanding of matter, the development of electricity, and the unveiling of the structure of the atom have led to the discovery of a new state of matter. It is more than 100 years since Crookes experimented with electrical discharges in vacuum tubes and suggested the existence of a new kind of gas: a gas composed of charged particles, such as a mixture of an electron gas and a proton gas. In 1923, the American chemist Irving Langmuir investigated the electrical discharges in gases and in 1929, while experimenting together with another American scientist, Levy Tonks, used the term plasma to describe the oscillations of the electron cloud during the discharge. This electron cloud was shining and wiggling, similarly to a jelly-like substance which reminded Langmuir of a blood plasma. However, the term plasma in Langmuir's experiments is completely misleading. Plasma in physics describes a gas of electrons and ions (atoms which have lost one or more electrons), while the plasma in blood is the clear yellowish fluid in which the blood cells are carried and is in a liquid form.

In the next chapter we will learn the important role played by the electrons in this new state of matter. We will also elaborate more on the importance of the ions and their interaction with the electrons. We will see the development of physics at very high temperatures for different densities and mixtures. We will see how chaotic and messy situations can become nice and orderly or vice versa. Using our imagination and everyday life experience, we shall lead the reader to learn about and get a feeling for what was originally the first, and today is the fourth state of matter—plasma.

Chapter 2

What is Plasma?

2.1 Introducing Plasma

Let's try to understand the four states of matter by calling upon our imagination. We are witnessing a dance competition. The conductor and the orchestra are ready to begin. The participants are well organized in pairs in a nice symmetrical way. This first phase is our solid—the competition has not yet begun and the atmosphere is cold. As the music begins and the pairs perform their first dance, a slow, we enter into our second phase—the liquid; the temperature is low as they dance to the soft music. The music picks up tempo. The dancers are doing the rock and roll and we enter the third phase—the gas; the temperature is getting warmer. Now the music blares as the pop tunes begin. The girls leave their partners and everyone is jumping and dancing by himself or herself. This is the last phase—the plasma; the temperature is very hot and everyone is jumping around all over the place. This example allows us to make an analogy in which the conductor is the physicist in the laboratory; the music is the 'heat' which changes the phases from solid, to liquid, to gas to plasma; and our dancers are the different particles of matter. The pair represent an atom (or molecule) which is the basic unit composing solids, liquids and gases. The girls represent electrons while their partners symbolize ions.

Although analogies may explain some issues, it is difficult to incorporate all the relevant facts. Therefore, analogies should not be taken too seriously. In particular, in the above situation, we have shown 'heat' (music) as an expression of motion, and atoms, electrons and ions (dancers) as the particles in the different phases. However, the interaction between the particles could not be directly expressed in the above analogy.

We have seen in Chapter 1 how plasma, originally the first state of matter, came to be known today as the fourth state of matter. When you heat up a solid (heating means putting energy into the system), it turns into a liquid. Further heating turns the liquid into a gas and still more heating gives us a plasma.

Let's be more specific and give an example. If you take a piece of metal such as iron and you heat it up to an appropriate temperature (more than 1000 degrees Celsius (°C) and less than 2000 °C) then this metal becomes a liquid just like ice turns to water at 0 °C (32 degrees Fahrenheit (°F)). If you heat this liquid metal further then it will become, at an appropriate temperature, a gas, just as liquid water becomes steam at 100 °C (212 °F).

Generally speaking we associate the four states of matter with different domains of temperature. As the temperature increases the substance passes from solid to liquid, through gas, to a plasma. The solid, the liquid and the gas are composed of a very large number of molecules (or atoms). In these phases of matter the heat is an expression of the *molecular* motion. In these three phases the molecules or atoms are moving as one unit. A plasma is composed of a very large number of free electrons and ions and in this phase of matter the heat is the expression of the *separate* electron and ion motions.

A gas is normally an electrical insulator, that is, electric currents cannot easily pass through it. However, by heating the gas to appropriate temperatures, physicists found that the insulator gas becomes a good electrical conductor. The gas is transformed into a plasma which consists of free electrons: the carriers of the electric currents. For example, money that is locked up in a safe is not as easily spendable as money that one carries in one's pocket. In this example the money represents the electrons that exist in atomic gas as well as in the plasma. However, whereas in an atomic gas the electrons are buried in the safe, the ones in a plasma (free electrons) are easily available for free motion. Owing to the free motion of the electrons a plasma is a good conductor of electricity. The plasma state is an electrically conducting medium. The medium contains electrons and positively charged ions and can also contain neutral atoms or molecules as well. Plasma will conduct electricity, like a wire of copper, because the free electrons can easily be moved.

A solid body resists compression as well as expansion. During the compression of a solid the repulsive forces between its molecules or atoms are dominant while during the expansion the attractive forces govern. This is why it is hard to squeeze or stretch a solid. The fact that the force between the molecules in a solid changes from one of repulsion at small distances (when squeezing) to one of attraction at larger distances (when stretching) suggests that there is an equilibrium state. When the solid is heated, the molecules start oscillating about their equilibrium position. As the temperature increases the vibrations of the molecules become more vigorous.

If the solid continues to acquire heat (i.e. energy) the vibrations of the molecules might become so large that the forces are unable to keep the molecules in their position any longer. When a large number of molecules leave their bound positions, the liquid stage is reached. The molecules in a liquid wander around inside. Occasionally some of the molecules from the liquid surface will escape and a vapor is formed. By further heating the liquid, one can cause more molecules to escape and transform into vapor. When the entire liquid is transformed into vapor the third state of matter, gas, is reached. Each molecule in the gas moves in a straight line with uniform velocity until it encounters other molecules (in a collision) or the walls of the vessel.

If the molecules in the gas acquire more energy they will first dissociate into the atoms forming the molecule. For example, a molecule of water will dissociate into hydrogen and oxygen atoms. Supplying the atoms with more energy causes the electrons to leave the atoms and to move freely inside the vessel. A plasma state is formed.

The forces that keep the solid in its structure as well as the forces that act on the molecules in the liquid are of an electrical nature. However, the electric forces in the plasma are stronger and more effective.

All material can be brought to a gaseous state by inserting the appropriate amount of energy. In a gas, the molecules or the atoms move freely inside the vessel, colliding mainly with the walls of the vessel. If the gas is not inside a vessel it diffuses into the air and eventually spreads throughout the atmosphere. If the energy is further increased, the molecules break up into their atomic constituents. What happens if we further increase the energy in our system? Because the atom has an internal structure, it will break into pieces just like the molecule broke before. If an atom loses one or more electrons, it becomes positively charged; this is called a positive ion. If one supplies the appropriate energy to the system the atoms will eventually break up into electrons and ions. Therefore, we see that by adding enough energy to any material, we can eventually produce a gas of electrons and ions. This last gas of electrons and ions is called a plasma (see figure 2.1).

Although plasma on Earth is man made, most of our Universe is in a plasma state. Our own Sun is in a plasma state and since almost all of the stars that we observe in the sky are suns, they, too, are in the plasma state. However, the planet Earth and all the other planets and their moons are not in a state of plasma. As more than 99% of the Universe is composed of plasma, it is vital to understand plasma physics in order to learn about the Universe and our surroundings.

Plasma physics is strongly related to applied sciences. In particular, international communications, energy power generators, and the nuclear fusion bomb (the hydrogen bomb) are examples of (good and bad) plasma applications. Nuclear fusion research is motivated by the need to solve the energy problem for generations to come. (This will be described in Chapter 5.) Today, there is an international effort to solve the energy problem and this has been the main stimulus for the development of laboratory plasma physics.

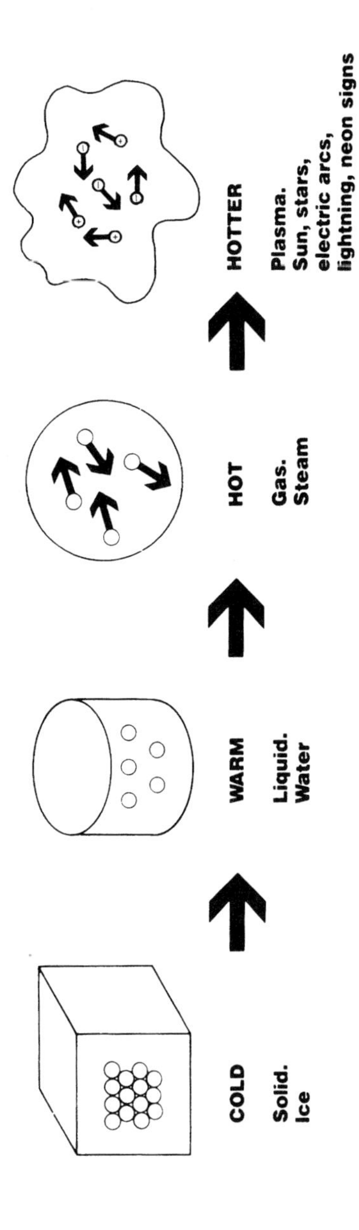

Figure 2.1 The four states of matter; plasma is sometimes referred to as the fourth state. (Courtesy of Lawrence Livermore National Laboratory, California, USA.)

2.2 A Visit to an Exotic Nightclub

If an atom or molecule loses one or more electrons, its total positive charge is greater than its total negative charge. The result is a positive ion. If the atom or molecule receives one or more surplus electrons, the total negative charge exceeds the total positive charge. The result is a negative ion. The flow of electricity through a solution or any liquid is caused by the motion of ions. The famous English scientist Faraday coined the word 'ion' from the Greek verb meaning 'to go'.

In order to explain the ionization process, i.e. the process in which electrons leave their atoms, we have to enter our imaginary phase and visit a very exotic nightclub. The main event is a striptease show done in the style of Eve, at a time when leaves were the latest fashion. The stripper's private parts are covered with leaves in such a way that before the show begins she is neutral—she neither attracts nor repels. She is therefore neither positive nor negative. She may look a little sexy, but attired in all those leaves, she does not attract too much male desire. However, each time she removes a leaf, she becomes more attractive and definitely more positive. By the time she finishes she is fully stripped. Scientists use the same term in science; when an atom has lost all its electrons, scientists claim that the atom is 'fully stripped'.

If we suppose that the stripper is an atom (with neutral charge) and the leaves are the negatively charged electrons, by throwing off the first leaf (the first electron) our stripper (or atom) becomes an ion. By throwing off two leaves, she is an ion with a degree of ionization of +2. Each time she removes a leaf, we call it the ionization process. When the atom is fully stripped it is called a nucleus (or the core) and is positive (as we are sure the audience would find our stripper). As long as some of the leaves are still on, the stripper can rearrange their positions. Our electrons can also jump from path to path. The minute she is fully stripped, the act of stripping is over. In physics, likewise, the ionization process is fully completed; however, the plasma game is just beginning.

Thus in physics the process by which ions are formed is called ionization. Ionization may take place in a number of ways. It can occur with the passage of radiation through matter or by heating matter to high temperatures. Heating the atoms means giving them energy. This energy is absorbed by the electrons and causes them to leave the atom. Therefore, we have a mixture of free electrons and whatever is left of the atom. Since each electron has a negative charge and every atom is initially neutral 'whatever is left of the atom' has a positive charge and is called a positive ion. In plasma, we are mostly dealing with positive ions. However, negative ions may also be formed if the freed electrons become attached to other atoms or molecules. The attachment will occur if energy is liberated when the electron collides with an atom or molecule.

Quasineutrality means almost neutral. For example, a system containing 100 protons (positively charged) and 101 electrons (negatively charged) is a quasineutral system. A plasma is usually a quasineutral system. The approximate equality of electron and ion concentration is due to the fact that the electric forces attract opposite charges. For example, if for some reason a bunch of electrons suddenly accumulate at a particular spot in a plasma, they will be immediately attracted by the ions, so that on average quasineutrality is satisfied.

Each piece of material is neutral. This is like your own bank account. For instance, you have just received your salary and are credited with this amount, which is the plus. Eventually, as the expenses accumulate, your account is debited. When you have spent all your income (which is the minus), you find yourself in a neutral state. Since we started with a neutral material (i.e. the gas which becomes a plasma), the plasma itself is neutral. However, as can happen with your budget, you might occasionally spend more than your income. In this case your balance does not come to zero. Similarly in a plasma, electrons can escape (remember the minus has left) and we are left with a slightly positive plasma; or, under some circumstances, one can 'earn' more electrons than ions so that a slightly negative plasma arises. This situation with slightly more negative or positive charges is what we mean by quasineutrality. Too much of the same charge put together will explode because of the force acting on these charges; just like too many minuses in the bank can cause bankruptcy. (Although too much saving does not fit in with our analogy; you can have many problems here as well.) However, very smart people know how to manipulate a lot of pluses (savings) as well as a lot of minuses (debts). You can still be a millionaire and owe a lot of money. In a similar way, very smart scientists can manipulate beams of positive charges and of negative charges (charged beams are bunches of electrons or ions).

In summary, we have seen that by giving enough energy to any material, it will eventually become a gas of electrons and ions. This last gas consisting of electrons and ions is called the plasma. The plasma moves under the action of electric forces inside a vessel. When plasma particles hit the vessel's wall, they are lost from the plasma system. Unlike a gas, the plasma cannot exist once its particles have collided with the wall. If the plasma is not in a vessel it diffuses into the air and spreads throughout the atmosphere, eventually becoming a gas as the ions and electrons recombine.

2.3 A Joint Ping-Pong Game

One person jogging is individual behavior. An American football game is collective behavior. For those who are not football fans, a performing orchestra is also a collective phenomenon. The charged particles inside a plasma are oscillating (swaying) collectively (jointly) in an orderly manner. For example, the electrons are moving in a collective way described by waves, usually called plasma waves. The plasma waves are a collective phenomenon.

Now let's analyze the collective situation in the plasma. The electron has a very small mass (weight) whereas the ion has a large mass. The ion is, therefore, much heavier than the electron. Thus, the electron will move much more easily than the ion (usually, a slim person can run much faster than a very heavy person). The electrons move together like waves in the sea. This wave motion of the electrons is a collective phenomenon. The ions, although moving more slowly, can also make waves with a smaller frequency. But what is frequency? If a particle is moving back and forth in space, then you count how many times it passes through a definite place. This is called frequency.

For example, let's take a ping-pong game. The ball is going back and forth. The number of times it reaches the bat per unit time (per second, per minute, per hour, etc) is called the frequency. In physics, scientists use the unit of seconds in their calculations. If two people play volleyball, their ball is much heavier than the ping-pong ball and therefore the frequency is smaller. In this example, the ping-pong ball is an electron and the heavy ball is an ion. Let's say that the ping-pong ball weighs one gram; then the ball used for the volleyball game should weigh a few kilograms in order to make our analogy sensible. However, ping-pong balls and volleyballs are not collective phenomena. They are not waves. In order to get a collective phenomenon, one has to describe a ping-pong game with millions of players simultaneously, so that when looking at the motion of the ping-pong balls one can see a wave motion. If we compare simultaneous games of electrons and ions (ping-pong balls and very heavy balls), it is clear that the frequency of the ping-pong balls is going to be much higher. Therefore, in a plasma one has collective phenomena, which are nothing else than waves of electrons and of ions such that the frequency of the electron wave is much higher than the frequency of the ion wave. Now if you think that plasma physics is only about waves describing very nice collective behavior, then you are wrong. Most physicists spend their time trying to understand plasma *instability* – why do the nice collective waves become unstable?

Let's return to the millions of ping-pong players who are playing simultaneously in a very nice and orderly manner. Imagine that one or a few players miss the ball or have a default. So what happens in this case? The other players don't pay any attention and the game continues. A few mistakes between millions of players are not significant. This is called a fluctuation. However, imagine that each player is somehow connected to the next player; that the default of one player causes the default of his neighbor; the default of his neighbor causes the default of his neighbor's neighbor and so on. In this case the game of the millions of players is crushed. In a similar way, a nice electron wave can be crushed.

Still another example of instability is watching a child build a pyramid out of ordinary playing cards. He builds and builds until a small fluctuation occurs (one card slips). The last card falls, causing the card placed before to fall and the one before that and soon the whole pyramid collapses. However, the pyramids in Egypt were not built in that way. They are stable. The aim of most scientists studying plasma physics is to make the collective behavior stable.

In a plasma state of equilibrium it is important to know what happens if one of the plasma parameters is slightly disturbed. If the disturbance grows, the plasma is unstable, while if the small disturbance decays and disappears the physicists say that the plasma is in a stable equilibrium. Plasma instabilities are classified into two large categories: (a) macroinstabilities—these are associated with a departure from equilibrium of a large part of the system; (b) microinstabilities—these are associated with small disturbances in the speed of the plasma particles which can increase and cause the plasma to be unstable. Physicists have investigated a variety of plasma instabilities. In this book we should not go into a detailed description of these instabilities. However, it is worth mentioning some of the colorful descriptive names given by scientists to plasma instabilities: 'sausage', 'kink', 'banana', 'firehose', 'flute', 'interchange', 'two-stream', 'tearing', 'ion-acoustic', 'loss-cone', 'mirror', 'Rayleigh–Taylor', etc, etc.

So far we have learned that an ionized gas in which all or a large number of atoms have lost one or several electrons and exist in a mixture of free electrons and positive ions is called a plasma. A standard definition of plasma is occasionally given as follows: a plasma is a quasineutral system of a large number of charged particles which exhibit collective behavior.

2.4 The One-Mile Run

How can we study millions and millions and millions of particles with many collisions moving in a chaotic motion? Physicists know, in principle, how to solve the trajectories of these particles. If one could track each individual electron and solve the equation of motion for this electron and every other electron, which means solving millions of millions of millions of trajectories (paths), then of course we would be able to know everything. But it is impossible for the human being to calculate and to understand what is going on by solving every particle separately. For example, take the population of a country or the population of the whole globe and ask what salary each individual earns or what is the standard of living of each individual. It is clear that writing down millions of millions of single persons' incomes is not feasible. It would take endless hours to write this down and the information obtained would be useless. For this purpose we use statistics to estimate the average income of an individual or the average standard of living in each country. In a similar way, physicists solve a problem for many, many, many particles by determining average quantities. The plasma is described by such quantities as temperature, pressure, density (number of particles per unit volume) and so on. It is meaningless to talk about the pressure or the temperature of one particle. Only the pressure and temperature of many particles has a meaning. The motion of many particles is calculated by using statistical mechanics. The main idea of this subject is to apply statistics when dealing with huge numbers. Thus the probability of different motions is calculated rather than the motion of each particle.

Statistical mechanics is the science of calculating the probabilities of things happening to moving particles. In a plasma there are free electrons and ions in motion. Are all the electrons moving at the same speed? Do they all have the same energy? The answer to both questions is no. The electrons do not move at the same speed and they do not have the same energy. The electrons move hectically inside the vessel, colliding with ions and other electrons. During and between the collisions some of the electrons and some of the ions lose energy. Some of the electrons gain speed and some of them lose speed. The theory of statistical mechanics calculates the energy distribution between the electrons which is described by a bell-shaped graph (see figure 2.2). In this figure N represents the number of electrons or ions, 'Energy' represents the energies of the particles and T is the temperature. This figure includes two bell-shaped graphs denoted by 1 and 2 representing two different distributions of the system. The temperature described by graph 2 is larger than that of graph 1. As one looks at the graphs, one can see what is happening. The top of the graph shows the energy shared by the largest number of particles. The bottom of the graph (bell) shows

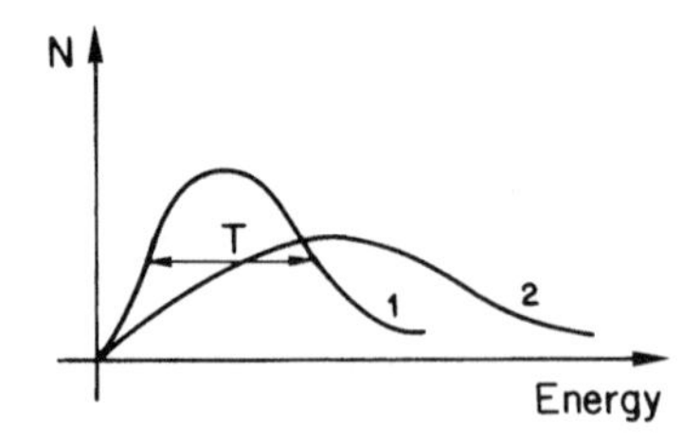

Figure 2.2 The energy distribution of electrons (or ions) in a plasma. Graph 2 represents a distribution with a higher temperature than graph 1.

that there are a few particles with very low energy and very few particles with very high energy. The width of the bell gives the temperature. The temperature is reflected in the spread of the energy distribution of the electrons (or ions).

Take, for instance, a container of gas and heat it up to a certain temperature. This temperature is measured with a thermometer. (As explained previously, the temperature is related to the motions of the atoms in the gas.) When this container is put inside a very fast moving vehicle, say a car, the thermometer will indicate the same temperature as in a stationary car. It is clear that the motion of the car does not add anything to the temperature of the container. (One can find oneself in a hot stationary car as well as in a cold moving car.) Therefore, for an outside observer and for one inside the car the temperature remains the same. However, the speeds of the molecules differ if measured by an observer inside the car and an observer outside a moving car. Thus care must be taken when relating the temperature to the speed (motion) of the molecules. This relation is properly defined at the center of mass which is the average position of the particles regardless of whether the vehicle is moving or stationary. In our example, the center of mass of the molecules in a container within the moving car is at a certain position inside the container. The temperature is proportional to the energy of motion relative to the center of mass and the energy of motion is proportional to the sum of the squares of the speeds of the molecules. The observer inside the moving vehicle calculates the gas temperature by measuring the speed of the molecules and adding their squares. The outside observer calculates the gas temperature by measuring the speeds of the molecules, subtracting the velocity of the vehicle from their speeds and then adding the resulting squares. This is summarized by the following:

(a) For an observer inside the car, the temperature is proportional to the sum of (molecule speeds) squared.
(b) For an observer outside the car, the temperature is proportional to the sum of (molecule speeds − car speed) squared.

Let's take many, many five-year-old children and ask all of them to run one mile. If we count the number of children for any definite speed, we get a bell-shaped graph. Very few children will be running very slowly; very few children will be running very quickly; while most of the children will be running at around the average speed. Now let's repeat the same one-mile run with the same number of children but this time with 15-year-olds. What will happen? We will get a similar graph. However, here the graph will become wider; the bell shape is wider and the top moves to the right which means that the average runner is moving faster. In this case the width of the bell-shaped distribution is wider and therefore, in the second case, we say that the temperature is higher (see figure 2.2, curve 2). The case of the electrons is slightly more complicated because they also collide with one another chaotically. The hotter the plasma, the greater the temperature, which means that the energy bell is wider. From this we learn that the temperature is related to the energy of motion of the electrons and the ions in the plasma.

In the above example, if all the children are running at exactly the same speed the temperature is zero. Why? Because there is no spread of energy. All of them have exactly the same mass (or so we assume for simplicity) and velocity (and therefore the same energy). If they run at different speeds, in order to calculate the temperature one has first to calculate their velocity relative to the center of mass. The average of the squares of these velocities (relative to the center of mass) is proportional to the temperature of the running children in our example.

Let's give another example to explain the center-of-mass motion. Take, for instance, a highway with cars speeding in one direction, say to the north. The cars are travelling with different velocities. For simplicity, we assume that all the cars have the same mass. In order to calculate the velocity of the center of mass we add the speed of all the cars moving north and divide the result by the number of cars. The result is the center-of-mass velocity. In order to calculate the speed of the cars relative to the center of mass, we have to subtract the center-of-mass speed from the velocity of each car. Once we know the velocity relative to the center of mass, we have to square the speeds in appropriate units in order to get the energy of the cars' motion with respect to the center of mass. This energy is defined as the temperature. This temperature turns out to be equal to the spread of the energy in our bell-shaped figure 2.2.

The distribution of the bell-shaped figure as a measure of the temperature of the particles was first calculated by the famous Scottish physicist James Clerk Maxwell in 1860. The same physicist was responsible for the final version of the famous equations describing the electromagnetic interactions associated with our famous agents the 'photons', which will be described in Sections 2.7 and 2.8.

Like temperature the density is also an important parameter of the plasma. The density is defined by the number of electrons or ions in a unit volume. Imagine that the unit volume is the size of a small box with dimensions of 1 cm in length, 1 cm in width and 1 cm in height. We can make an analogy by picturing a big elevator with one or two people. In this case we would say that the density is very low. Taking the same elevator with 20 people inside makes the density higher. However, placing a few hundred people in the same elevator makes it very crowded and the density grows significantly higher. In a plasma, we can have from a few particles to millions of millions of billions of particles in a unit volume.

Different plasmas are classified according to their temperature and density. Laboratory plasma can acquire different densities and temperatures depending on the different devices used. A variety of densities and temperatures exists in the plasma of the atmosphere, the Sun, the stars and the Universe.

The density and temperature are related to the pressure of the plasma. Pressure of a plasma is induced by the motion of its particles and is measured as the force applied per unit area. In plasma the pressure is usually calculated by

$$\text{Pressure} = \text{Density} \times \text{Temperature}.$$

From the above one can see that the pressure increases with increasing temperature or density.

In this section we have given an outline description of the plasma in terms of statistical mechanics, pointing out the importance of such terms as probabilities and average motions. A plasma is described by average quantities such as temperature, density and pressure. In the next section we introduce another plasma parameter related to the nature of the electric force which 'rules' the plasma.

2.5 Shielding

Usually between friends there is an attraction while between enemies there is a repulsion. We find in nature particles that attract one another (like friends do) and those that repel one another (like enemies do). Opposite charges (+, −) attract one another while similar charges (+, + or −, −) repel one another. This is due to the electric interaction between charged particles.

The force between charged particles is described by the Coulomb force. This force is named after the French physicist Charles Augustin de Coulomb who discovered this law towards the end of the 18th century. The Coulomb force is inversely proportional to the square of the distance between the charges. For example, if between + and − there is a distance of one centimeter (cm), and the force of attraction is one unit of force, then if one places the two charges at a distance of 2 cm the force will be a quarter of the previous force (i.e. 1/4 of the unit of force). When they are placed 10 cm apart the force equals 1/100th of the previous unit of force. On the other hand, if the two charges are placed only half a centimeter apart, the force of attraction between them is 4 units. This force is also proportional to the charged quantity of the particles. (The unit of measuring charge is called the coulomb.) The larger the charges, the larger is the force.

Coulomb's law, which describes the forces between charged particles, extends to infinity. For example, if the distance is very, very large, the force is not zero although it is very, very small. An electron will feel the force of an ion no matter what the distance is between them. This situation is modified in a plasma, although the basic law that describes the forces between the charges is the Coulomb law.

Imagine a positive ion in a plasma. A cloud of electrons is attracted around this positive ion due to the Coulomb law. The electrons closest to the ion form a shield for the other electrons in such a way that the force between the shielded electrons and the ion is smaller than the Coulomb force without the shielding. Such a shielding also exists for electrons further away and the force becomes gradually weaker for each layer. Therefore, the force of attraction of a positive ion does not extend to infinity but to a finite distance. This distance is called the Debye radius or Debye length and the shielding is called Debye shielding, after the Dutch physicist Peter Debye who explained the motion of ions in solutions. The forces of repulsion between ions and independently between electrons are also reduced by Debye shielding.

The Debye radius is a function of temperature and density. For a hotter temperature the Debye radius becomes larger. For a denser plasma the Debye radius becomes smaller. As the density is increased (unit volume of plasma contains more electrons), the shielding is more effective, and the Debye length is decreased. On the other hand, for a higher temperature, the motion is greater and the shielding is thus less effective, causing the Debye length to be increased. The temperature causes the plasma pressure to increase, which in turn compels the plasma to expand. The Coulomb force tends to keep the plasma together. Therefore, without thermal motion (no pressure) the charge cloud around an ion would collapse to an extremely small value.

The ionization of only a few atoms does not create a plasma. A gas is characterized as a plasma if the density of its electrons is large enough that the Debye length is smaller than the dimensions of the plasma system.

2.6 Collisions

In order to distinguish between the different phases of matter it is useful to understand the concept of 'mean free path'. This is defined as the average distance travelled by a given particle (that is, a typical particle whose motion we follow) between collisions with the medium in which it moves. The mean free path is inversely proportional to the density of the medium, that is, for higher density the mean free path becomes smaller. From this we learn that since the density of the solid phase is usually larger than that of the liquid and the liquid density is larger than that of the gas, the mean free path of a particle varies in each phase. In a solid the mean free path is usually smaller than in a liquid and in a liquid it is much smaller than in a gas. In the plasma state of matter, one can find densities similar to those in all the different phases. How can the properties of the plasma and the other phases be distinguished when comparing their mean free paths? This can be done by noting that the mean free path is inversely proportional not only to the density of the medium, but also to the probability that the given particle will collide with other particles in the medium. This probability of collision is described by the term known as cross section (measured as an effective area).

Let's imagine two bowling games being played with two different sized balls: small balls and large balls. The balls are rolled down a long narrow lane to knock down a group of pins (pins much narrower than are used in a regular bowling game). It is obvious that the large balls can knock down more pins than the smaller sized balls, assuming that the player is capable of using both balls skilfully. Splitting a ball in half enables us to see the cross section of the ball, namely, the area of the flat surface of the cut ball. The bigger ball has a larger probability of knocking down more pins.

In order to estimate the probability of a collision between two balls, one has to measure their cross sections. It is clear that the cross section of a collision between two large balls is bigger than that between two small balls or one large and one small.

When dealing with charged particles one also has to take into account the strong and long-range influences of the electric forces. In this case the cross section between two colliding charged particles is determined not by their geometrical areas but rather by the effective area of the electric forces. The latter is usually much larger than their geometrical area.

Furthermore, due to the nature of the interactions between the particles in general, the cross section is usually dependent on the relative velocity between the colliding particles. In particular, for the Coulomb interaction, the cross section decreases rapidly with increasing relative velocity. This is because as the particle moves faster, the probability of its being deflected by the collision is smaller.

Let's pretend that we are watching children playing a game called 'catch'. A certain area is designated in the middle of the field where a number of children are waiting to catch the 'intruder' whose aim is to cross over to the other side of the field. The probability that he will be caught is much larger if he runs slowly and much smaller if he runs quickly. This probability also depends on the number of 'catchers'. The analogy that we make here is the following:

Number of 'catchers' ↔ interaction strength
Speed of the 'intruder' ↔ relative speed of the particles
Area where the 'catchers' are moving ↔ cross section.

Thus we can see how the probability of the event under consideration (described by the physicists as a cross section) depends on the interaction strength as well as on the relative velocity between the particles.

Table 2.1 Mean free path for different densities in the four phases of matter.

	Density	Cross section	Mean free path
Solid	Very large	Large/medium	Very small
Liquid	Large	Medium	Small
Gas	Small	Small	Large
Plasma	Various	Very large	Various

The above discussion of cross section enables us to distinguish between the mean free path of two different phases with the same density. For example, consider a gas and a plasma with the same density (i.e. the same number of particles per unit volume). The constituents of the gas phase are neutral molecules while the constituents of the plasma are charged particles. Therefore the cross section for collision in the first case (the ordinary gas) is just like the case of the collision between balls; it is given by their geometrical dimensions. In the second case (the plasma) the cross section is much bigger owing to the Coulomb forces between the charged particles of the plasma; it depends on their effective interaction areas. For larger cross sections, the mean free path is smaller. For a gas and a plasma of the same densities, the mean free path in a plasma is much smaller than that in a gas. The properties of the mean free path are summarized in tables 2.1 and 2.2.

Another important feature in the collision phenomenon is the collision frequency. This describes the number of collisions of a given particle in a unit of time (say in one second). The collision frequency is inversely proportional to the mean free path, i.e. for longer mean free paths the collision frequency is smaller (there are fewer collisions in a second). The collision frequency also depends on the speed of the particles. From the dependence of the mean free path on the density and on the cross section for collision, the collision frequency is related to the quantities

$$\text{Collision frequency} = \text{Density} \times \text{Cross section} \times \text{Relative speed.}$$

The above relation is of general importance in physics and can be applied for counting the rate of events in any physical process. Since in a plasma the cross section for collisions decreases significantly with increasing

Table 2.2 Comparison between a gas and a plasma.

	Density	Cross section	Mean free path
Gas	Same	Small	Large
Plasma	Same	Very large	Small

Table 2.3 A comparison between the collision frequencies of plasma particles and regular balls.

	Density	Cross section	Collision frequency
Regular balls	Same	Geometrical area (which does not depend on speed)	Increases with increasing speed
Plasma	Same	Decreases with increasing speed (i.e. increasing temperature)	Decreases with increasing speed (i.e. increasing temperature)

speed, the collision frequency decreases with increasing speed. In a closed vessel the motion of the charged particles is related to the temperature. In a plasma, the collision frequency decreases for increasing temperatures. Thus for higher and higher temperatures, the collision frequency becomes smaller and smaller, and for extremely high temperatures the plasma becomes effectively collisionless. The collision frequency of charged particles of the plasma relative to that of regular 'balls' (such as billiard balls) is given in table 2.3.

2.7 Swallowing and Ejecting Photons

Atomic physics is the science of the electrons in the atom. Nuclear physics is the science of the protons and the neutrons inside the nucleus. The electrons tend to move in a circular motion around the nucleus similar to the planets moving around the Sun. This motion is described by quantum mechanics (very bizarre effects and incomprehensible to our ordinary everyday way of life). Although quantum mechanics is also important in the physics of very dense plasmas, we shall not discuss this subject in this book but shall concentrate on classical science.

The electrons move around the nucleus because they are bound to it by electric forces. Remember that the minus (electrons) is attracted to the plus (nucleus). In order to pull the electron out of its path, we have to invest energy. This energy can be supplied by a photon or by a collision with another electron.

In 1905 Albert Einstein introduced the concept of the photon. A photon is an uncharged particle, has zero mass and moves with the speed of light. A photon is a quantum (packet) of radiant energy. Photons can cause excitation of atoms and molecules and more energetic ones can cause ionization.

When a photon collides with an electron (see figure 2.3(a)) the electron jumps to a higher trajectory (further from the nucleus). When a bound electron collides with a free electron, the bound electron can jump to a

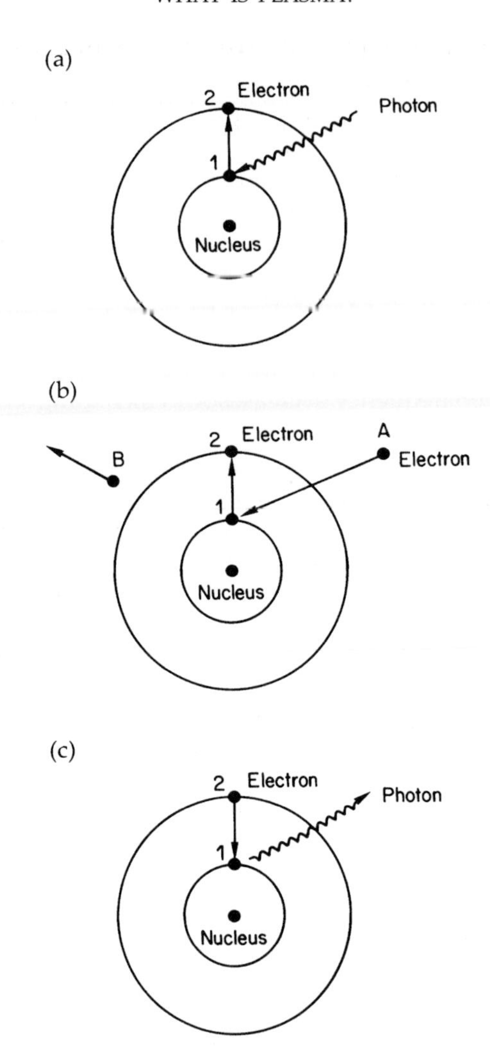

Figure 2.3 (a) A 'bound electron'–'photon' collision causing a change in the electron trajectory. (b) A 'bound electron'–'free electron' collision causing the bound electron to jump to a higher trajectory. (c) The 'bound electron' jumps to a lower trajectory by ejecting a photon.

higher trajectory (figure 2.3(b)). When the electron jumps back to its normal path, it emits a photon (figure 2.3(c)). It's like swallowing a ball and ejecting it. An electron can swallow a photon but not an electron (since an electron can swallow a photon and still remain an electron with the same charge and mass).

In a plasma, we have millions of millions of millions of free electrons and ions and neutral atoms. Some of the ions and neutral atoms have electrons spinning round. Photons and colliding electrons in the plasma can cause these bound electrons to jump from path to path, resulting in the creation of more photons in the plasma. These photons are part of the electromagnetic radiation inside the plasma. Occasionally, some of these photons escape from the plasma causing loss of energy. A hot plasma is cooled by the loss of energetic photons. The photons that leave the plasma in our Sun heat our planet. 'Sunshine' is a flux of photons arriving from the Sun.

2.8 The Agents

A charged particle creates an electric field in the surrounding space. A moving charged particle creates a magnetic field as well. In general, physicists claim that moving charges create 'electric + magnetic = electromagnetic' fields. Since plasma consists of moving charged particles (electrons and ions) one finds electromagnetic fields in a plasma.

Now what is a field? A field can be represented pictorially by a set of lines of force. The density of these lines at any given point in space represents the strength of the field at that point. The direction of the lines represents the direction of the force acting on a particle at that position. In order to understand this pictorial view of the field, let's look at 'Newton's apple'. Between two masses there is a gravitational force of attraction, that is, the apple is attracted by the planet Earth. The force of gravity emerges from the Earth in all directions and is transferred from the planet Earth to the apple. Here comes the pictorial line of force to explain this problem. One can imagine that a set of lines of force hits the apple. The apple feels the force and is attracted to the Earth. In a similar way to the gravitational force, there are forces between charged particles. As a matter of fact, scientists have shown that the electrical force between charged elementary particles is far stronger than the gravitational force. Every force can have a few 'sources' that create it, and each 'source' can be related to a field. However, for simplicity, we describe each force by one field. The gravitational force has the gravitational field, the electromagnetic force has the electromagnetic field, the strong nuclear force has its own strong nuclear field and the weak force has its own weak field. Similarly, each different sport, such as baseball, football, basketball and so on, has its own field for each appropriate game.

Each source sends agents to create the field. For example, the charged particles send out 'photons' (particles of light) as their agents to accomplish this task. The photons are not charged particles but their task is to interact with other charged particles that they come in contact with. These photons are thus the agents of the electromagnetic field. This can be compared with an agency center (the source, e.g. the charge) sending

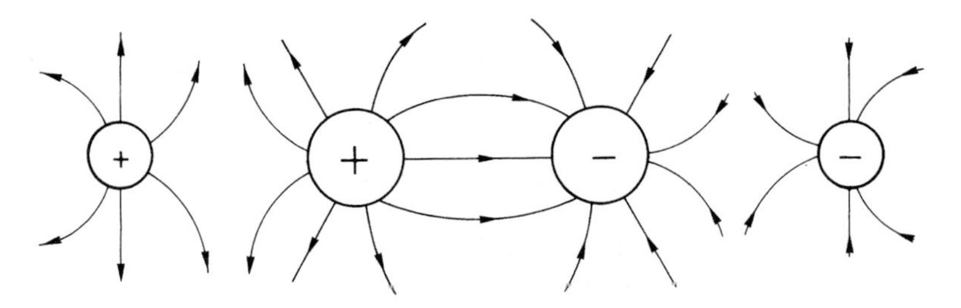

Figure 2.4 The electric field created by positive and negative charged particles.

out agents (the photons or equivalently the field) everywhere, interacting with other agents and sources.

Between charges there is a force so that opposite charges (+, −) attract one another (like opposite sexes), while between similar charges (+, + or −, −) there is a repulsive force. It should be noted that this is not the case for the gravitation force between masses. The masses have a positive sign and they still attract one another. As was previously said, the forces are associated with an imaginary field. For a positively charged particle, one imagines lines of force radiating outwards from the positive charge. Therefore, if there are two positive charges, each one of them with outward lines of force, one gets the picture that the lines of force push against one another. Therefore, positive charges repel one another. A negative charge such as an electron also has lines of force coming towards it. In a situation with one positive charge and one negative charge, the fields do not collide, but rather the charged particles attract one another (see figure 2.4).

We already mentioned that a moving charged particle creates a magnetic field. But what is a magnetic field? What is a magnetic force? To answer these questions, we have to remember that there are ordinary magnets that are capable of attracting nails or screws. Such a magnet has a 'north pole' and a 'south pole'. If we bring two magnets together in such a way that the two poles are facing each other, then a force is created between the magnets. This is called magnetic force. Opposite poles (north, south) attract one another and similar poles (north, north or south, south) repel one another. Thus far we have a similar situation as with the electric charges. The force between magnets was also explained by Coulomb in 1785. This force, like the other forces, can be described by a field, the so-called magnetic field. In this field picture, we can imagine lines of force going out from the north pole and coming back into the south pole (see figure 2.5).

One already sees a difference between electric and magnetic charges. In a magnet the field lines are always in the form of a closed loop and a

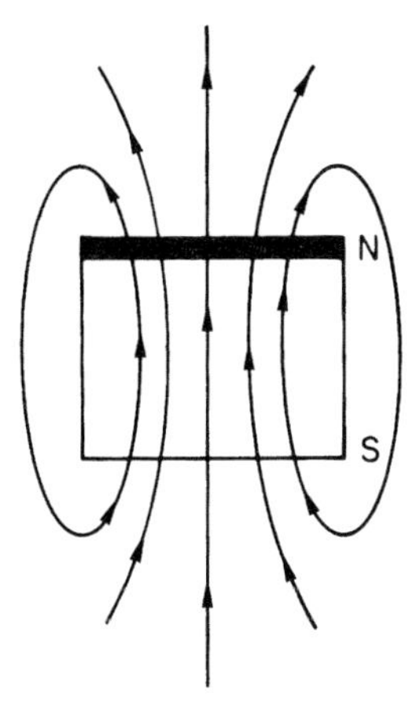

Figure 2.5 The magnetic field created by a magnet; N denotes the north pole and S the south pole.

picture describing the magnetic field created by a single 'magnetic charge' cannot be plotted. Whereas for the electric charge the force is an individualist (see figure 2.4), the magnetic force is always described by a pair (north, south); see figure 2.5. One can cut the magnet in half, and those halves in half and the new halves again in half, and so on, and each piece will be a magnet with two poles, north and south. However, as we already know, by dividing any matter into smaller and smaller pieces, we can eventually get a single electron with a negative charge. This, of course, is not possible for a magnet, since physicists have not so far found 'magnetic charges'.

Besides the permanent magnets that are capable of attracting nails and other 'iron' objects, there are electromagnets which possess the same property. An electromagnet is a coil of wire wound around a central core (or alternatively an appropriate arrangement of copper bars). When an electric current is passing through these coils (or the bars) they become like permanent magnets. A magnetic field is introduced into the vicinity of these coils (or copper bars) by the electric current flow. A loose piece of iron in the vicinity of the electromagnet will be attracted by the coil (or copper bars). Around the permanent magnet, or the current-driven magnet (electromagnet), there are lines of magnetic force that are capable of attracting metals like iron. These lines of force can actually be visualized by the following experiment. Sprinkle some iron powder on a plastic plate. A wire (or bar of copper) is placed beneath the plate and is connected to an electric circuit. When the current is switched on the iron powder will arrange itself in patterns of curving lines exactly like the magnetic lines of force.

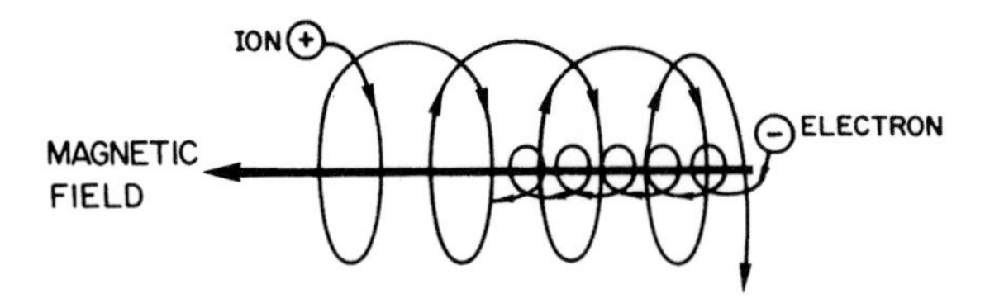

Figure 2.6 Motion of an electron and an ion in a magnetic field.

The magnetic lines in physics are analogous to contour lines in geographic maps. In considering a map with contour lines one knows that the terrain is steep if the lines are close together while the land is flat and slopes are gentle if the contour lines are far apart. Similarly the magnetic field is strong (and attracts pieces of iron very powerfully) if the magnetic field lines are close together, and weak if the magnetic lines are separated.

You are probably wondering by now what all this has to do with plasma. The answer is very much. Plasma is a bunch of moving charged particles of electrons and ions. Each electron or ion moves along a magnetic field line and is confined to the vicinity of this line until it is disturbed (see figure 2.6). The electrons and ions move in spiral trajectories around the magnetic field lines in such a way that usually the electron gyration radius is smaller than that of the ion.

Let's return to the moving charges and the magnetic fields. A moving charge creates an electric current. As explained above, electric currents create a magnetic field like the field created by a magnet. The direction of this magnetic field follows the 'right-hand rule' (see figure 2.7). This rule may be visualized when a light bulb is screwed into the socket. The screw thread is in the direction of the current while the person's thumb points in the direction of the magnetic field that the current produces. This magnetic field applies a force on other charged particles. As you can see, this is a complicated, continuous cycle:

motion of charges ↔ magnetic forces ↔ motion of charges ↔ etc.

Let's visit an average household. During an ordinary day, at one point or another, the child becomes very boisterous and loud. His noise annoys the father, who cannot concentrate on his reading. He scolds the child. The child begins to cry. This in turn irritates the mother, who in turn snaps at the father. The poor father has to escape to another place to read his newspaper in peace, but soon after, his child finds the same place for his boisterous game and the cycle is once more repeated.

Returning to physics, a moving electric particle creates a magnetic field, which in turn applies a force on other moving charged particles, which affects the motion of the original moving particle, and so on and so on. Since in a plasma there are many, many, many charged particles, the

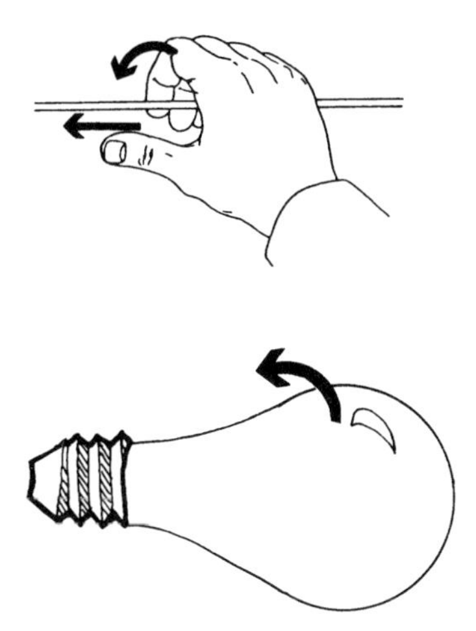

Figure 2.7 The 'right-hand rule' that shows the direction of magnetic field lines that are created by an electric current. The 'right-hand rule' is a convenient way to remember how magnetic fields are oriented with respect to the currents which produce them. When your thumb points in the direction of the current in the wire, your fingers curl the way the magnetic field goes around the current. When your fingers are curled in the direction of the current in a coil, your thumb points in the direction of the magnetic field it produces. (Courtesy of Lawrence Livermore National Laboratory, California, USA.)

situation is far from simple and is occasionally very messy. This statement is also correct when using mathematical formulations.

Speaking 'physics' is using mathematical equations. In the language of physics one writes equations to describe the motion of charged particles due to the forces interacting between the particles. In an equivalent way, one can write equations describing the motion of the charged particles in the electric and magnetic fields. There are two equivalent descriptions: that of forces and that of fields. Usually, it is more convenient to describe the motion of the charged particles using the description of fields. As this book is not written in 'physics' but rather in simple 'English', the mathematical equations are omitted, although, on some occasions, they are very simple.

2.9 Safekeeping

Conservation laws are also used to calculate the motion of electrons and ions in a plasma. In physics many things are conserved. In particular, energy is a conserved quantity. The energy changes from one kind to another: from heat energy to motion energy, from heat energy to electric energy, from nuclear energy to heat energy, etc, etc. However, the total energy of the system is conserved. It is like a quantity of money moving from person to person, through the bank, through different safes. Let's start with one million dollars. When counting and adding all the amounts together, the total remains the same. No matter how many hands the money has passed through, one can always count to the million dollars. The same is true of the energy flowing inside the plasma. The energy of all the electrons and ions together does not change, although the distribution and circumstances can alter. Energy is not the only quantity conserved in a plasma system. The mass is also conserved. The plasma can move but the weight remains the same. It is like a glass of water. When you pour the water from one glass to another, the weight of the water is the same, assuming that you don't drink or spill any of it.

Take for instance two pounds of strawberries and one pound of sugar. If you put them in a dish inside the refrigerator, even after one week the weight will remain the same. On the other hand, if you make a jam out of the strawberries and sugar by cooking this mixture, many atoms will escape at the time of the cooking. When the jam is fully cooked, the mixture weighs much less. If one could collect all the atoms that escaped during the cooking, one would have the same weight after the cooking as before. This conservation seems self-evident when making jam; however, it is amazing how much one can learn from the conservation laws when cooking a plasma.

Complementary to the conservation of mass and energy, a strange quantity called 'momentum' (defined as the mass times the speed) is also preserved. The rate of change of momentum is proportional to the force acting on a particle. This is Newton's famous law of motion for a particle as well as for a 'piece' of plasma. In the plasma, the force is given by the action of electric and magnetic fields. From these self-evident conservation laws, the physicists know how the plasma is going to move and what is going to happen. The understanding of the motion of a plasma is important for the confinement of the plasma inside a vessel.

2.10 Plasma Reflections

(a) Plasma in solids

Solid-state electrical conductors, such as copper, silver and many other materials, possess electrons that can move freely inside the conductor. Because of the free motion of these electrons we can use copper wires to transfer electricity from power stations into our homes. The 'free' electrons are not completely free; they interact with one another due to Coulomb forces. Moreover, as the solid is neutral the negatively charged electrons are balanced by the positively charged ions. Thus the motion of the electrons is occurring in a background of positively charged particles, where the ions are fixed and the electrons are moving. This is similar to what we considered in the plasma case. It was thus concluded that collective oscillations of the electrons occur in a solid. Therefore, the understanding of plasma is important for a better understanding of the solid state. The electron density, i.e. the number of 'free' electrons in a cubic centimeter of a solid material, is very high. These plasma electrons also possess quantum effects and thus a discussion of quantum plasma physics is necessary, which is outside the scope of this book.

(b) Non-neutral plasmas

From our discussions on plasma, we have learned that the plasma is quasineutral; that is, although locally the plasma need not be neutral the overall plasma system is neutral. However, this condition is not an essential requirement. One can define a non-neutral plasma as a collection of charged particles in which one does not have an overall charge neutrality. Physicists have proved that non-neutral plasmas exhibit collective properties that are similar to those of neutral plasmas. For example, powerful beams of electrons exhibit collective oscillations and other plasma properties. The oscillating electrons in tubes that generate microwaves also follow the plasma behavior. The problems of equilibrium and instabilities of intense electron beams (or other charged beams such as protons) confined by magnetic fields are considered to be similar to the neutral plasma confinement in magnetic fields. The physicists involved in studying conventional elementary particles with large accelerators realized that the theories and the computer simulations of plasmas are very useful in understanding and analysing the physics of big accelerators.

(c) Free-electron lasers

The free-electron laser device consists of an electron beam moving at velocities close to the speed of light in a magnetic field. The magnetic field is arranged in such a way that the electrons that are passing through are deflected alternately in opposite directions. For example, imagine a set of magnets arranged in a row with opposite polarities. The electron beam passes first between the magnets of the north pole and then between the magnets of the south pole; and, alternately through north–south–north, etc poles. This combination of magnets is called a wiggler. In this way the electrons execute a wiggly motion through the magnetic field and therefore radiate photons at the frequency of their oscillations. (An oscillating electron radiates electromagnetic waves.)

Taking into account the electron velocities, the wavelength is significantly shorter. This effect is similar to the sound of a whistle of a moving train. As the train moves towards you, the sound is higher in pitch than that of a stationary train. The faster the train moves towards you, the higher is the sound; as the train speeds towards you, the sound waves become more closely spaced and the tone becomes higher. This change in tone or frequency is called the Doppler effect, named after the Austrian physicist Christian Johann Doppler who first explained this effect correctly in 1842. In a similar way, radiation is Doppler shifted to shorter wavelengths due to the fast speed of the electrons. For more energetic electrons, the wavelength of the radiation becomes shorter. Present-day devices produce laser radiation at wavelengths from the region of millimeters to the regions of fractions of a micrometer (visible light). Today new ideas and developments are being considered which may further reduce the wavelength of the radiation. X-ray lasers may be developed in the future from such a device.

The physics of free-electron lasers is the physics of non-neutral plasmas. The science of intense electron beams moving in periodic magnetic fields is the science of plasma. A better understanding of the physics and technologies involved is needed in order to obtain efficiently shorter- and shorter-wavelength lasers.

(d) Electron–positron plasma

A positron is an electron with a positive charge and is the antiparticle of the electron. The positron does not exist naturally on our planet. It 'arrives' in cosmic radiation or is created in the laboratory. When a high-energy intense electron beam hits a target with a high atomic number such as tungsten, positron fluxes are created. Such experiments were first attempted at Cornell University in the USA to create electron–positron plasmas in the laboratory. Although such plasmas can be confined for only milliseconds (the collision of an electron with a positron causes the annihilation of the electron–positron pair and two or three photons are created), it is a long enough time for very interesting and exotic plasma experiments.

The main physical motivation for electron–positron plasma research stems from the existence of this type of plasma in stars such as the white dwarfs and pulsars which are thought to be rotating neutron stars. The electron–positron plasma is a typical example of a particle–antiparticle plasma system. Such particle–antiparticle plasmas may be of importance in the understanding of extremely dense stars, such as neutron stars. Moreover, these plasmas might be developed in the laboratory as interesting sources of radiation.

2.11 Plasma Compendium

In summary, plasma is a medium containing many charged particles governed by electromagnetic forces. Plasma is a quasineutral gas of charged particles which possess collective behavior. Plasma is the term used in physics to designate the fourth state of matter; matter that exists in the Sun and the stars, in space, etc. Most of the Universe is made up of plasma. The plasma medium is described macroscopically (on a large scale) by its temperature and density, and changes in the plasma are calculated by using conservation equations such as conservation of energy, momentum, and mass. On a microscopic (small) scale the plasma is described statistically using probabilities for calculating the positions and the velocities of all particles. Due to their mutual collisions, the plasma charged particles emit radiation (electromagnetic waves). Moreover, there are many different waves that can be created in a

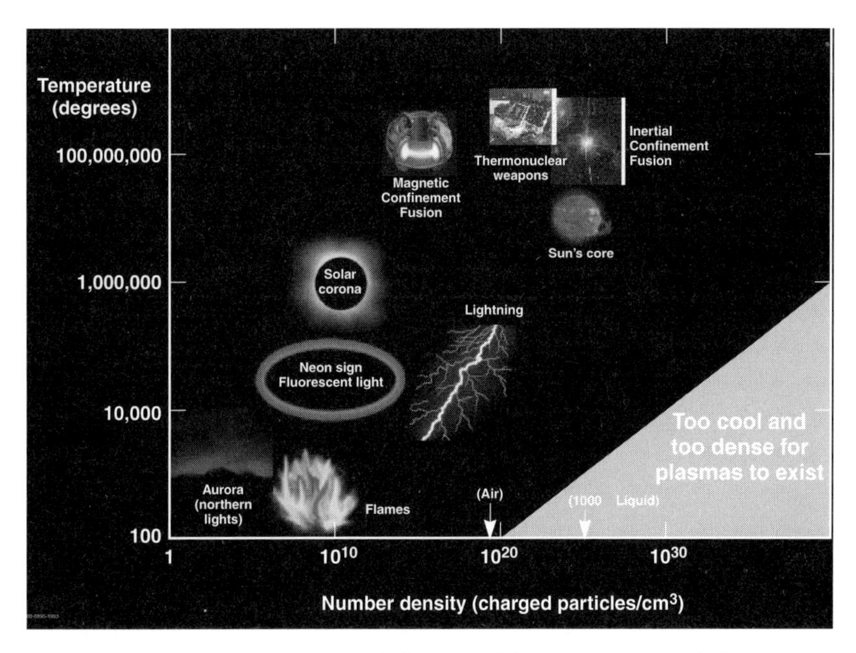

Figure 2.8 Temperatures and density of the Fourth State of Matter. (Courtesy of the University of California, Lawrence Livermore National Laboratory and the US Department of Energy.)

plasma medium. The stabilities and instabilities of these waves play a crucial role in plasma systems.

Plasma is essential in explaining the Universe (see Chapter 3); plasma is extensively used in industry to improve the standard of living (see Chapter 4); and scientists believe that the understanding of plasma can lead to the solution of the energy problem for all future generations (see Chapter 5). Throughout the book the history of plasma is intertwined with the explanations of the plasma data and phenomena. More about the history of plasma is given in Chapter 6.

Chapter 3

A Universe of Plasma

3.1 Plasma in the Beginning

The Belgian astronomer Georges Edward Lemaitre suggested in 1927 that in the beginning (zero time) all the matter and energy of the Universe was squeezed together into a 'cosmic egg'. This cosmic egg was not very stable and therefore it exploded. This explosion was the most gigantic of all time; so big that it is very hard to imagine its real force and vigor. The fragments of the explosion became the galaxies and the stars of today. This model was popularized and further developed scientifically by the Russian-born American physicist George Gamow. Gamow called this model the Big Bang theory. In this model the Universe is expanding and the galaxies are receding from each other. Independently of the Lemaitre and Gamow model, the American astronomer Edwin Powel Hubble concluded in 1929 that the galaxies were receding from one another.

Let's imagine the Universe as the surface of a balloon on which certain areas represent stars and galaxies. Between the stars is the interstellar medium. If one blows up the balloon, it expands and the distance between the stars increases. A similar picture, although geometrically more complicated, describes our Universe.

There is no scientific evidence for the origin of this cosmic egg or what happened before the explosion. The scientific evidence begins at an extremely short time after the explosion. At that time, in the beginning, everything was in a plasma state of matter. This plasma state was very bizarre and different from the one which was discussed in the previous chapter. It consisted of extremely high-energy photons and charged elementary particles. Due to the many collisions between the photons and the charged particles, as well as between the particles themselves, the photons maintained the same temperature as the other particles. The temperature was so hot that the radiation photons had enough energy to produce elementary particles. In order to understand the formation of particles in this extremely hot plasma, one needs some basic knowledge of elementary particles and the laws of conservation that govern their interactions.

For example, it is a well established fact that electric charge is a conserved quantity. Photons do not have any charge, while the electron, for example, has a negative charge. When photons combine and produce an electron, with a negative charge, another particle with a positive charge has to be created so that the charge of the whole system is still zero. The two photons have zero charge, and therefore the two particles that are produced from the photons should have a sum of zero charge. Thus two photons can create an electron and a positron, the latter being exactly like an electron but with a positive charge. The positron is also called the antiparticle of the electron. All particles have their own antiparticle partner. For example, the proton, with a positive charge, has a partner called the antiproton which has the same weight as the proton but with a negative charge. The electron, the proton and the neutron, which are the building blocks of matter, have antiparticle partners, the positron, the antiproton and the antineutron, which are the building blocks of antimatter. Our Universe is built from matter but it is possible that it also contains antimatter stars, antimatter galaxies, etc, with their appropriate antielements. An antimatter star contains a plasma of antiprotons and antielectrons (which are the positrons). When matter and antimatter collide, they annihilate each other and transform into energy consisting of photons and a collection of particles and antiparticles. For example, the collision of an electron and a positron will create two or three photons.

During the first stage, the Universe was so hot that particles and antiparticles were created easily from the radiation. The photons were very energetic so that this could be achieved. For example, two photons creating an electron and a positron should each have an energy at least as large as the mass of the electron. Energy is also a conserved quantity. Energy can change from one form to another; from mechanical energy to electrical energy; from nuclear energy into energy of heat; from radiation energy into particle energy and so on, and vice versa. Energy can never be derived from nothing. The conservation of energy means that the quantity of energy in the Universe that was available in the 'cosmic egg' is exactly the energy available today in our Universe and this quantity of energy is constant.

During the extremely hot period in the first stages of the Universe, most of the plasma was made up of electrons and positrons, neutrinos and antineutrinos. There were also protons and neutrons but in less abundance. Neutrinos have very bizzare properties. So far it is believed that they have no mass and thus move with the speed of light, like photons. They have, however, spin, similar to that of electrons. They do not possess any electric charge and are not affected by electromagnetic interactions. They only interact with other particles through the weak interaction, which is related to beta decay. The existence of neutrinos has been established experimentally. However, their mass might differ slightly from zero, but so far this issue is an open question. By the end of the millennium, new experimental evidence, still not completely conclusive, is suggesting that the neutrinos possess a non-zero mass. The non-zero mass was theoretically suggested by and published in 1974 by S. Eliezer.

During this extremely hot period there existed a plasma of protons, electrons, positrons, neutrons, neutrinos, antineutrinos, and photons and the most important nuclear reactions at that stage were as follows:

$$\text{Antineutrino} + \text{Proton} \rightleftarrows \text{Positron} + \text{Neutron}$$

$$\text{Neutrino} + \text{Neutron} \rightleftarrows \text{Electron} + \text{Proton}.$$

However, at this stage, the numbers of protons and neutrons were almost equal. Complex nuclei had not yet formed.

The temperature of this Universe was dropping rapidly (the Universe was expanding and cooling) and the ratio between the numbers of neutrons and protons changed. For lower temperatures neutrons could not be produced as quickly as protons in the above reactions. Moreover, since the free neutron is not a stable particle it decays and disappears.

The neutron decays into a proton plus an electron and an antineutrino. The lifetime of the free neutron is about 15 minutes. However, the neutron is bound in stable nuclei. Its lifetime inside the nucleus is infinite, so that the neutron is stable in this case. A neutron inside an oxygen nucleus, for example, does not disintegrate like a free neutron.

As the plasma cooled, the neutrinos became free particles and left the hot matter. At this stage the electrons and positrons annihilated each other by creating photons. The temperature was still hot enough for protons and neutrons to fuse and form deuterium. A further nuclear fusion between deuterium and a proton formed a helium-3 nucleus, which fused with a neutron to create a very stable helium-4, consisting of two protons and two neutrons. The development of the Big Bang model predicts that the Universe consists of about 25% helium and 75% hydrogen. (The free neutrons decayed and the only neutrons left at the beginning were the bound ones inside the helium.) There is very little helium here on Earth; the reason is that the helium atoms are light and they do not easily interact with other elements by chemical reactions and therefore escaped from our planet a long time ago. Helium was actually first discovered in 1868 by J. Norman Lockyer when analysing the spectrum of our Sun. Astronomical observations of our Sun and other stars and galaxies are consistent with this prediction of the Big Bang model, namely, that the stars in the sky (the plasma suns) consist of about 25% helium and 75% hydrogen.

The most significant confirmation of the Big Bang model was the discovery of the temperature of the background radiation of our Universe. This radiation has been measured in a uniform way all through space where no other stars or galaxies are shining. It is thus concluded that this radiation is the 'left-over' photons from the beginning of our Universe.

Scientists were able to calculate the properties of these 'left-over' photons from the Big Bang. The Universe cooled down to a very low temperature, about three degrees above absolute zero (i.e. minus 270 °C beneath the freezing point of water). These experimental measurements were made in 1965 by the Bell Telephone Laboratory radio astronomers Arno A. Penzias and Robert W. Wilson. Their measurements are direct support for the Big Bang model of the Universe.

Let's bring back our 'Universe balloon' depicting the expansion. Looking at the places between the representations of stars and galaxies we suddenly notice some low-energy photons in a uniform distribution. As these 'empty' spaces are not 'inhabited' by stars and galaxies which produce photons, the question is asked where do these photons come from? Scientists believe that these 'left-over' photons are the remains of the Big Bang event, when the 'balloon' was extremely small and contained many photons.

To summarize, there are three well established facts about our Universe which support the Big Bang model: (a) Hubble's important discovery of the receding galaxies; (b) the composition of the stars, consisting of 75% hydrogen and 25% helium; and (c) the 'three degrees above zero' background radiation of the Universe as measured by Penzias and Wilson. All these facts are consistent with geologists' estimates for the age of our planet Earth which they believe to be about 10 billion years (a billion equals one thousand million). At present there is no other theory which explains the development of the Universe that is consistent with the above established facts. For this reason scientists today believe that the Big Bang theory correctly describes the development of our Universe. In this theory, plasma was the first and only state of matter at the beginning of our Universe.

3.2 The Universe

Science in the beginning was mostly based on speculation. Hundreds of years ago men believed that our planet Earth was the center of our Universe and that the Sun, the Moon and other stars moved around our planet once every day.

Today, we know that our planet Earth moves around our Sun. The Sun's 'family', also referred to as the solar system, includes nine planets: Mercury, Venus, Earth, Mars, Jupiter, Saturn, Uranus, Neptune and Pluto. Some of the planets have more than one moon (for example, Jupiter, which is the largest planet, has more than 12 moons). The planets, their moons and other various bodies, such as comets and meteorites, scattered throughout our solar system, all move around our Sun. The Sun and its solar system are travelling in space.

Five of the planets, Mercury, Venus, Mars, Jupiter and Saturn, are visible at night to the naked eye. These visible planets, as well as our Moon and other 'star-like' objects in our solar system, do not have a luminescence of their own. They are seen by reflected sunlight since they are like mirrors that reflect light. The planets and their moons are not in a plasma state of matter. However, the other stars that are seen at night are in plasma form. Thus, the fourth state of matter dominates our sky.

Our Sun belongs to a collection of stars consisting of about one hundred thousand millions of suns. This gigantic system is called a galaxy. Our Sun belongs to the galaxy called the 'Milky Way' which is only a minor part of our Universe. Through the use of sophisticated telescopes, astronomers are able to view thousands of millions of galaxies. All these galaxies are in a plasma state of matter. Our Sun, while spinning on its own axis, is travelling around the center of its galaxy. The galaxies are receding from one another and our Universe is expanding and is becoming larger and larger.

Our Sun is a medium-sized star. It is very close to us, relative to the other suns, and therefore it appears much larger and hotter than the other stars. The distance between our Sun and the planet Earth is huge (by our everyday standards), but on the astronomical scale, it appears that the Sun and the planet Earth are very, very close indeed. Our Sun and its family are just a dot in the Universe.

Since the distances between the stars are so immense, it is convenient to measure the distance between them by the time that light takes to traverse these distances. The speed of light, the fastest possible way to transfer information from one place to another, is 300 000 kilometers per second. This is a huge number, but still a finite one. In order to get an idea about the speed of light, it is useful to remember that a pulse of light leaving our Sun will reach the Earth after 8 minutes and 20 seconds! The astronomical distance between the Earth and our Sun is therefore 8 light minutes and 20 light seconds while the distance between the Earth and the Moon is only one and a quarter light seconds. Thus, in a year, light will travel about 10 thousand billion kilometers. In order to have a better understanding about these large distances, astronomers call the distance that light can travel during one year a light year. The second nearest sun to us is a faint star in the sky known by the name of Alpha Centauri which is about 4 light years away from us. The light from it that we see when looking through a telescope actually left Alpha Centauri four years ago.

The closest galactic neighbor to our Galaxy (the Milky Way) is the Magellanic Clouds, at a distance of about 170 000 light years. There are two galactic systems in this structure of stars: the so-called 'Large Cloud' and the 'Small Cloud'. This cloud-like condensation of stars is visible to the unaided eye in the southern hemisphere.

The next nearest galaxy to the Milky Way is the Andromeda Galaxy. As its distance from us is more than two million light years, we see the Andromeda Galaxy as it used to be long before the beginning of civilization on the planet Earth. Although Andromeda seems to be so far away, it is quite close to us on an astronomical scale. If one knows the exact position of this galaxy in the sky, one can see it with the naked eye.

Each galaxy consists of hundreds of billions of suns and there are billions of galaxies. The galaxies are not uniformly distributed and form groups called clusters. The Milky Way, of a spiral shape, is a member of a cluster which includes 30 galaxies. The largest galaxy in this group is Andromeda. An average galaxy contains hundreds of billions of stars and a typical size (diameter) is 100 000 light years. Not all the suns are alike, but all of them are composed of plasma. Even the intergalactic matter (the matter between the galaxies) is in a plasma state. Thus, plasma is the state of matter of most of our Universe.

The motion of our Universe is dictated primarily by the laws of gravitation. As most of the Universe is composed of plasma, its structure is due to a merger between the gravitational force and plasma properties. For example, a star is in equilibrium due to the combined actions of the plasma pressure (determined by the fusion reactions in the star) which tends to expand it and the gravitational force which tends to collapse it. While plasma physics was developed only during the 18th century, the laws of gravitation were introduced by the outstanding English scientist Sir Isaac Newton in the 17th century. Moreover, Newton's early research on light and color contributed significantly to the later observations and understanding of the stellar bodies.

Not all the plasmas in the stars have the same temperature. Not all the stars have the same luminosity. In many stars, the energy output that they radiate changes with time. These variable stars play a special role in astronomy. In particular, the cepheids and the novae are extremely important because they make it possible to establish the distances of remote systems beyond our Galaxy. Cepheids are pulsating stars that change their radiation in a periodic way, while novae change their luminosity in an explosive way. The explosive novae are divided into two groups: (a) classical novae, where the explosion involves only the outer layers of the stars; and (b) supernovae, where the explosions are catastrophic. Usually, novae are small stars which are much fainter but hotter than our Sun. During the explosion, the stars brighten suddenly and fade again, usually within a few hours. Supernovae explode and in a very short period of time become a hundred million times brighter than our Sun, and then fade away. The cepheids, the novae and supernovae are all in a plasma state of matter.

Most of the stars in the Universe are members of binary systems, revolving one around another. About one-half of all the stars in our

Galaxy (the Milky Way) are binary stars. It appears that our own Sun is an exception as no companion is known. The binary stars play an important role in calculating the masses of stars. The masses are deduced from the measured trajectories of each star and its companion.

The stars differ from one another in luminosity, temperature, size and density. In particular, plasma properties are defined by the temperature and density of the matter (see Chapter 2). There is a sequence of stars with increasingly large densities: (a) white dwarfs; (b) neutron stars; and (c) black holes. White dwarfs have a mass comparable to that of our Sun but with a density about 100 000 times greater than that of water! The plasma properties of the above very dense stars are very unusual and possess quantum plasma effects which are not discussed in this book.

A class of stars which is even denser than the white dwarfs is the neutron stars. In order to comprehend their densities one has to imagine, for example, a situation where the whole mass of our Sun is squeezed into a sphere with a radius of about 15 kilometers. A neutron star is so dense that it is effectively a single huge nucleus; one cubic centimeter of this star weighs hundreds of millions of tons. The extreme pressures and temperatures in these stars cause the electrons and the protons (the ion of hydrogen is a proton) of the plasma to fuse into neutrons; hence the name neutron stars. These stars were discovered in the 1960s and were called pulsars because they were found to be emitting strong rhythmic optical and radio pulses.

The densest stars of all are black holes. The gravitational force in these stars is so powerful that neither matter nor light can escape from them. The irradiated light from this unusual plasma is trapped in the black hole and thus these stars are not directly seen. Moreover, these stars act like a 'bottomless pit', as all matter that enters their gravitational sphere is swallowed up.

The stars are changing and their birth, development and death can be described by astronomers. Although the exact picture is not known in detail, the following scenario is accepted today. Stars are formed from the grains, the gas and the plasma of the interstellar medium (the space between the stars and between the galaxies). Occasionally, a massive cloud will form and then the gravitational force of this cloud will condense it further (see figure 3.1). As this plasma cloud is further increased to high enough densities, the temperature rises to many millions of degrees, causing thermonuclear reactions which supply the energy of the star. Usually over billions of years the hydrogen fuel in the core of the star is converted into helium by thermonuclear reactions. There are also other possible nuclear fusion reactions inside the star which change its structure. This period of nuclear burning is the stage of development. The way and the time that the star is going to die depend mainly on its mass, and whether it has a companion (binary star). Stars can collapse into white dwarfs or

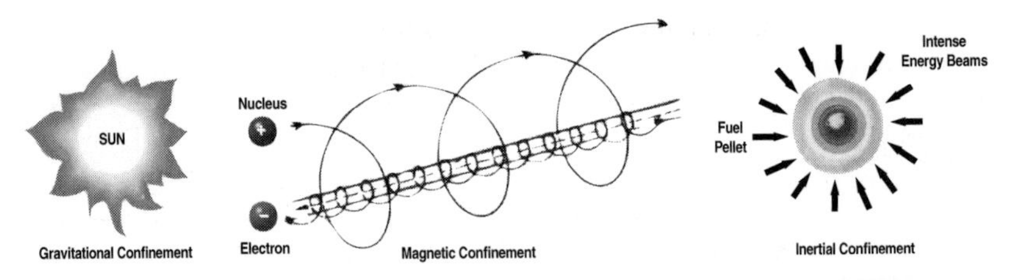

Figure 3.1 The three types of possible confinement: gravitational, magnetic and inertial. (Courtesy of Lawrence Livermore National Laboratory, California, USA.)

neutron stars or black holes. This is the last stage—their death. Our own Sun will die as a white dwarf after becoming a giant star. The Earth will probably be swallowed up during this 'giant' stage. However, we do not have to worry now about this catastrophic end of our planet because this astronomical picture will not occur for thousands of millions of years.

3.3 The Magnetosphere

Most of the matter in the Universe today is in the form of plasma. Magnetic fields exist throughout the Universe. In astrophysics, research into plasma in magnetic fields has become an important issue. The plasma parameters, such as the density and temperature, and the magnetic fields in the Universe cover a wide range of values. For example, the number of plasma particles in a cubic centimeter is about one in the interstellar medium (space between the stars), about one million in the outer region of the Sun, about many million billion billion in the interior of the Sun and other stars in our Universe and even more in white dwarfs and neutron stars. The temperature in the interstellar medium is about 10 000 degrees, about one hundred times hotter in the outer region of the Sun (the corona) and many thousand times hotter in the interior of the stars. The values of the magnetic fields also change drastically from the intergalactic plasma (very low magnetic fields) to the plasma of white dwarfs and neutron stars (very high magnetic fields).

Plasma is continuously flowing from the surface of the stars under the influence of magnetic fields. In particular, plasma emerges from our Sun into interstellar space; this flow is known as the solar wind (containing protons and electrons). This flux of plasma encounters the Earth's magnetic field in its path. Some high-energy particles of the solar wind are caught and confined by the Earth's magnetic field.

The Earth has a significant magnetic field along its surface. This magnetic field decreases as the distance from the Earth increases. The magnetosphere is the region in the atmosphere where magnetic fields affect the motion of

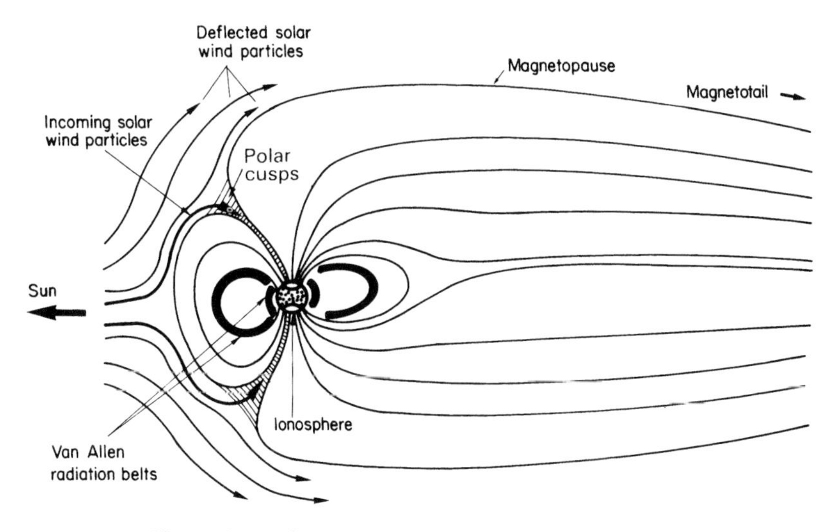

Figure 3.2 The magnetosphere of the planet Earth.

the plasma charged particles arriving from the Sun. At a distance of 10 times the Earth's radius (at about 65 000 kilometers) in the direction of the Sun, the pressure of the solar wind is equal to that of the Earth's magnetic pressure. This equilibrium region is called the magnetopause (see figure 3.2) and it defines the outer boundary of the Earth's magnetosphere. The lower boundary is a few hundred kilometers above the Earth's surface. In the direction away from the Sun, the night side, the Earth's magnetic forces and the solar wind are parallel and therefore the magnetosphere extends to a much longer distance. The magnetosphere on the 'night' side is a few hundred times larger than on the 'day' side (see figure 3.2). It is convenient to measure distances in the solar system by astronomical units, usually denoted by AU. One AU is the average distance between the Earth and the Sun, which is about 150 million kilometers. The magnetosphere of the Earth in the 'night' zone is as large as a few AU. The Earth's magnetosphere can be described as being in the shape of a comet with the Earth located at the head of the comet and the magnetospheric 'tail' as being on the opposite side, away from the Sun. The magnetosphere can also be pictured as a cylindrical shell with a blunt nose.

One of the most famous domains in the magnetosphere is the Van Allen radiation belts (see figure 3.2). In 1958, Van Allen discovered two doughnut-shaped zones of charged particles, such as electrons and protons, which are trapped by the Earth's magnetic field. The charged particles are 'caught' by the magnetic force of our planet Earth and follow corkscrew motions around these magnetic lines of force. The charged particles are confined in the Van Allen belts exactly like a plasma is confined in a magnetic mirror (see Chapter 5). The electrons (or protons) spiral about

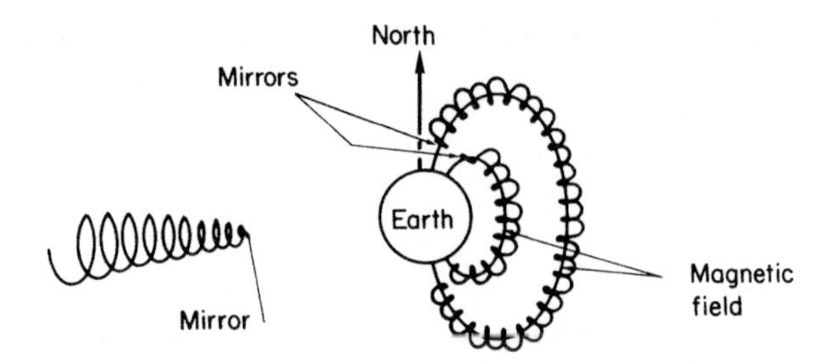

Figure 3.3 The 'magnetic mirror' configuration of the Van Allen radiation belts.

the Earth's magnetic lines of force in such a way that when approaching the north or south regions, where the field is stronger, the spiral path becomes smaller and smaller until the charged particles reverse their direction of motion (see figure 3.3). The Earth's magnetic field has a configuration like that of a regular magnet. This configuration of magnetic fields is called a dipole. The magnetic force near the North and South Poles is larger than that around the Equator (as the lines of force are closest near the poles). The magnetic fields at the North and South Poles act like mirrors for the charged particles in the Van Allen radiation belts (see figure 3.3). There are two Van Allen radiation belts, the inner and outer belts, differing in the energy and composition of the charged particles. While the inner belt consists mainly of high-energy protons, the outer belt is dominated by electrons. Beneath the inner belt there is a region of cold plasma of electrons and protons. This plasma is in equilibrium with the Earth's ionosphere; this is the region of plasma that acts as a mirror for radio waves. The propagation and reflection of waves in the ionospheric plasma is of vital importance to radio communications.

Another interesting and spectacular plasma phenomenon close to Earth is the luminous aurora. Aurorae, dancing columns of light in a variety of forms, are caused by electrical discharges in the upper atmosphere. Around the North and South Poles, one can see different shapes of light formations, such as rays and dancing arcs. This light is emitted from plasmas formed by fast electrons arriving from the Sun and caught near the magnetic poles of our planet Earth. The electrons collide with the atoms of the atmosphere and ionize them. While the maintenance of the ionosphere by photons from the Sun is a permanent phenomenon, bursts of fast electrons from the Sun that create the aurorae are the result of Sun storms and other disturbances. Therefore, the ionosphere is permanent while the aurora is a transient phenomenon. The 'solar cusps' in figure 3.2 are regions between the magnetic lines of force in the northern and southern hemispheres that allow the solar-wind plasma to enter and produce the day-side portions of the auroral zones.

The magnetosphere also contains a variety of plasma waves which cover a wide range of frequencies. For example, during auroral displays, the plasma radiates waves with wavelengths of the order of kilometers. These waves are produced above the ionosphere at an altitude of about the Earth's radius (about 6400 kilometers).

Professor Hannes Alfven explains its origin as a result of the existence of plasma double layers. Although plasma is usually referred to as a quasineutral gas, sometimes an accumulation of two layers of positive and negative charges occurs. This double layer is described by two oppositely charged layers inside the plasma that are separated by a distance of the order of the Debye length (see Chapter 2). This separation of the plasma charges is very significant. Very strong electric fields are associated with such a separation of charges. The charged particles of the plasma, or other cosmic charged particles, can be accelerated by these double layers. They are also associated with electrical current flow through the plasma. In particular, currents run from the magnetosphere to the ionosphere along the magnetic field lines. Instabilities in such plasmas with their double layers and electrical currents are believed to generate radio waves during the auroral displays. Double layers and electric currents flowing from one domain to another appear to exist in the plasma magnetosphere. The interstellar medium of our Sun and perhaps other regions in our Universe are like huge 'electronic circuits' in a plasma medium. It is interesting to point out that plasma double layers have also been observed in laboratory experiments for a wide range of plasma temperatures and densities. Although Irving Langmuir discussed the existence of double layers in his first publication in 1928 on plasma, today they are still under investigation.

The planet Earth is not the only planet with a strong magnetic field. Through the unmanned *Mariner Venus–Mercury* spacecraft scientists have found that the planet Mercury (which is a little larger than our Moon) has an intrinsic magnetic field which is about a hundred times smaller than that of the Earth. This magnetic field creates Mercury's magnetosphere. This magnetosphere extends to one planetary radius above the surface of Mercury. Figure 3.4 describes schematically the dimensions of the magnetospheres of the planets.

As one can see from this figure, the largest planet Jupiter has the largest magnetosphere (in relative and absolute terms) due to its magnetic field, which is about 10 times larger than that of the Earth. This magnetosphere extends towards the Sun for about a hundred Jupiter radii and is at least a hundred times larger in the anti-Sunward direction. In order to comprehend the huge size of Jupiter's magnetosphere, it is interesting to point out that the spacecraft *Voyager 2*, which was traveling with an average speed of 35 000 kilometers per hour, took 32 days to cross only a portion of this magnetosphere.

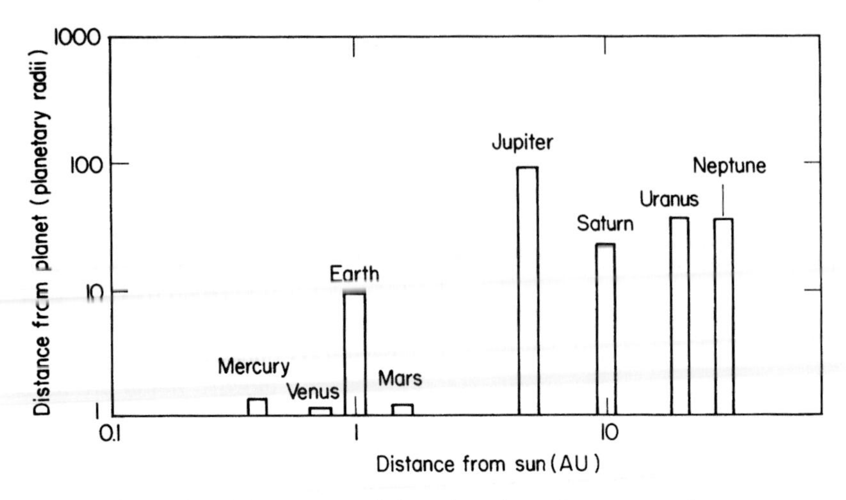

Figure 3.4 Dimensions of the solar planets' magnetospheres.

Measurements from the spacecraft *Pioneer 11* showed that the planet Saturn has a magnetic field comparable in relative terms with that of our planet. The planets Venus and Mars have been studied by many Soviet and American spacecraft and no significant intrinsic magnetic fields have been found. The very small magnetosphere of Venus seems to be caused by the direct interaction between the solar wind and the ionosphere of Venus.

Exploration by spacecraft is continuing and the different magnetospheres of the solar system are under continuous research. Scientists believe that research into plasma in our solar system is important and indeed crucial in understanding the plasma phenomenon in our Universe.

3.4 Light From the Stars

Our knowledge of the Universe has been deduced almost entirely from measurements of photons. These photons are electromagnetic waves arriving from the stars and measured by telescopes and other instruments. The electromagnetic waves have two oscillating components at right angles: one component is the electric field and the other is the magnetic field. The two components are propagating in the perpendicular direction.

Until about 50 years ago, astronomers could detect only photons of visible light (red, blue, etc). As different devices were developed photons with less energy than those of visible light (such as infrared and radio waves) were measured. With the use of high-altitude balloons and the development of artificial satellites, scientists were able to extend these measurements to include photons which cannot penetrate our atmosphere.

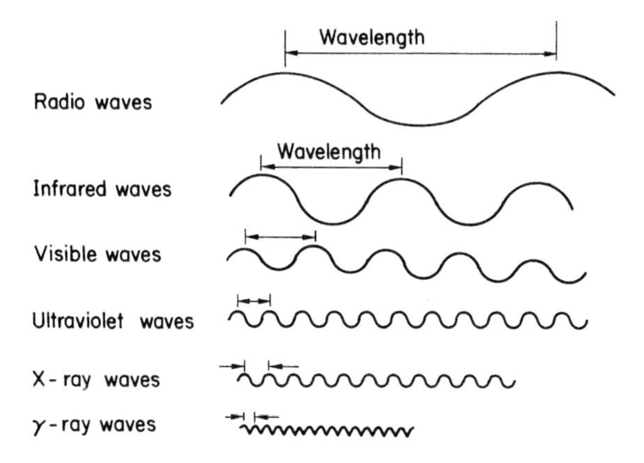

Figure 3.5 A schematic description of the wavelengths of the electromagnetic spectrum. γ is the Greek letter gamma.

Artificial satellites are capable of measuring photons of light defined as ultraviolet, X-rays and gamma rays. The gamma rays are the most energetic photons (see figure 3.5). The wavelengths in figure 3.5 are schematic and not drawn to the right proportional scale. For example, a typical radio wavelength, which can be as long as one kilometer or more, can include one hundred million infrared waves. In one infrared wavelength only 10 to a hundred visible waves can be inserted. The ratios between the wavelengths of gamma ray and X-ray, X-ray and ultraviolet, and ultraviolet and visible are typically of the order of one-tenth. The energy of the photons is inversely proportional to their wavelengths, i.e. if the wavelength is a hundred times smaller the photon's energy is a hundred times larger.

In 1666 Sir Isaac Newton analyzed a beam of sunlight that entered a darkened room through an opening in a window blind. The beam of light passed through a prism (a triangular piece of glass). The light bent while passing through the prism and shone on the wall as a rainbow. In this simple experiment, Newton showed that the white visible light arriving from the Sun contains red, orange, yellow, green, blue and violet components. Newton called this rainbow band of colours a spectrum. The white light from the Sun spread into a rainbow because each color bent at a different angle while passing through the prism. Although we see sunlight as a white light, it is actually a mixture of many colors, as was shown by Newton's experiment.

In 1800 the famous German–English astronomer William Herschel studied the temperature appearing on a thermometer when exposed to different regions of the spectrum of sunlight. In this way he found the energy content of the light of each color arriving from the Sun that was split through the prism. To Herschel's surprise, he saw a rise in temperature on the thermometer *beyond* the red end of the visible spectrum. There was no visible light there but nevertheless energy had arrived from the Sun. From this experiment Herschel suggested that sunlight included wavelengths of light longer than any that could be seen and sensed by our eyes. He called this light 'infrared', meaning 'below the red'. This was the first discovery of an invisible light. Today we know that there exist other invisible lights such as the radio waves which have even larger wavelengths and the X-rays and gamma rays which have much shorter wavelengths. Visible light is only a very small portion of the electromagnetic wave spectrum that exists in nature. The spectrum that is produced by our Sun and by the other stars consists not only of the red-to-violet colors that we can see but also of invisible regions of light below the red and above the violet. The difficulty of detecting shorter wavelengths than the violet, namely the ultraviolet, the X-rays and the gamma rays, in astronomical observations is due to the fact that these electromagnetic short waves cannot penetrate our atmosphere. These short waves were detected only after the advent of high balloons and artificial satellites.

The spectrum of a Sun, or any other object, will change according to its temperature. For example, when heating a piece of iron to high temperatures we will first see it turn a red color. If the temperature continues to rise, the iron is seen to take on an orange color, and then a whitish color at higher temperatures. As the temperature increases, the main wavelength of radiation becomes shorter. For example, from the Sun's spectrum scientists know that the temperature of the external surface of the Sun—the so-called photosphere—is about 6000 °C. The Sun's interior is much hotter and reaches temperatures of about 10 million degrees Celsius. It is known that both the very hot interior of the stars and their less hot photospheres are made of plasma. The physics of the Sun's interior as well as of its exterior is the physics of plasma.

The light from the stars is not just a continuous wavelength form, as described by Newton's rainbow of colors, but includes also 'lines' of light which have a specific and well defined wavelength (or energy). The German physicist Gustav Robert Kirchhoff and the German chemist Robert Wilhelm Bunsen had shown in 1859 that each element produces its own characteristic spectrum. With increasing temperature, the electrons surrounding the atomic nuclei move from a lower trajectory to a higher one, that is from a lower energy state to a higher one. After a specific lifetime in the excited state, the electrons jump back to the original state while emitting a photon of light. Since the electrons can jump from trajectory to trajectory in a number of ways (see figure 3.6), a particular type of atom will emit (or absorb while jumping up) a number of different photons with different wavelengths. These photons are recorded on a special photographic plate in such a way that every wavelength arrives as a line at a different position on the plate. This set of lines is called the

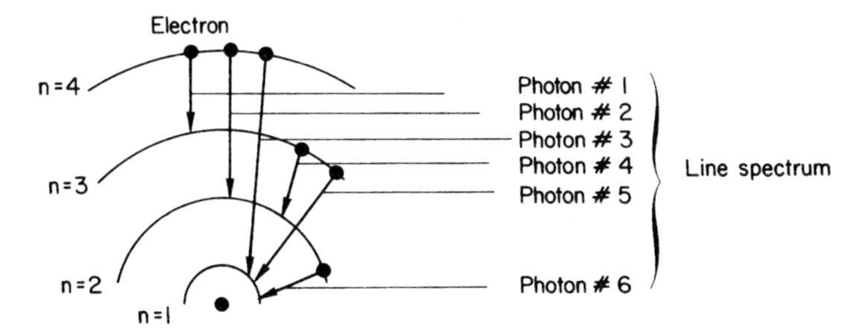

Figure 3.6 A schematic description of the creation of a 'line spectrum'.

'spectral lines'. Every element in Mendeleyev's Periodic Table has a different number and arrangement of electrons and a distinctive spectral pattern. For example, an iron atom, which has 26 electrons, produces thousands of lines in the visible spectrum and many more other lines outside the visible range of light. It is for this reason that spectral lines can be used to identify the elements here on Earth as well as those on our Sun or on any other star. The emission-line spectra of the elements are their 'fingerprints'. No two elements produce lines in precisely the same positions on the photographic plate, just as no two people can have the same fingerprints. In this way the element helium was discovered first on the Sun before it was found on Earth. From these spectral lines scientists are able to deduce not only the temperature of the stars but also their composition. In 1814, the German optician Joseph von Fraunhofer was the first to detect the 'missing lines' (i.e. absorption of photons) in the Sun's spectrum. Today scientists are constantly trying to detect and to analyze the spectrum of highly ionized atoms inside a plasma state of matter in the laboratory as well as in stellar plasmas. Since the temperature in the stars is high enough for ionization, the contents of our Universe are thus seen through a 'plasma spectrum'.

3.5 The Star's Interior

The Sun's core sustains a temperature nearly a hundred thousand times hotter than that of the Death Valley desert in the USA. The pressure is about 10 million times greater than at the bottom of the Pacific Ocean. The solar center is composed mainly of hydrogen and helium matter. Under the above extreme pressures and temperatures matter is compressed to a density about 150 times larger than that of water. Energy is produced by thermonuclear fusion at the center of the Sun.

Ten tons of 'Sun material' is needed to produce a rate of energy of about 2 watts, so that 500 tons of Sun plasma would be required to light a 100 watt bulb. Although the nuclear fusion energy released is extremely large the process of energy production is very inefficient. The huge quantity of energy released by the Sun is due to its enormous size. The ratio of power produced (power = energy rate) to weight of the Sun is less than the same ratio for a burning match. It would therefore be useless to solve our energy problems by the same thermonuclear fusion that occurs naturally in the Sun. As we shall see in Chapter 5 there are fusion processes which are very efficient. In particular, a one meter sphere filled with deuterium–tritium plasma at a temperature and a density comparable to that in the center of the Sun would radiate as much energy per unit time as the entire Sun itself!

The Sun's inefficient energy production is crucial for the survival of mankind. If the rate of energy production in the Sun were as efficient as the deuterium–tritium fusion process, the Sun would have burnt out long before the beginning of human civilization. It is interesting to point out also that even if the Sun's energy production were a little more efficient (say 10 times today's efficiency, which would still be extremely small) then the Earth's temperature would be so hot that no civilization would be possible.

The core of the Sun, referred to as its furnace, emits huge quantities of energy due to its extreme size of about one million miles in diameter. Our Sun's diameter is a hundred times larger than that of the Earth and its volume can include more than a million planets the size of Earth. The energy produced in the interior of the Sun radiates from the deep interior towards its surface and further into space. This energy transport is occurring in a plasma medium. The equations governing the structure of the Sun, its energy production and its energy transport are solved by the use of big and fast computers. The understanding of plasma physics plays a vital role for formulating and solving these problems.

Let us discuss in more detail the processes that produce energy inside the furnace of the stars. The German-born American physicist Hans Bethe suggested in 1938 a sequence of thermonuclear reactions explaining solar energy production by the so-called 'carbon cycle'. A small quantity of carbon inside the Sun plays a very important role in producing its energy. The carbon nucleus fuses with the hydrogen yielding the isotope of nitrogen ^{13}N (see figure 3.7). During this fusion process an energetic photon—a gamma ray—is liberated. The nucleus of ^{13}N is unstable and therefore it decays into a stable isotope of carbon, ^{13}C. During the decay, a positron (an elementary particle possessing the same mass as the electron but of opposite charge) is emitted together with a neutrino. The isotope of carbon ^{13}C fuses with a proton to produce the stable nitrogen ^{14}N. ^{14}N fuses with another proton to form an unstable isotope of oxygen, ^{15}O, and another gamma photon. ^{15}O decays into an isotope of nitrogen, ^{15}N, while emitting a positron and a neutrino. Finally, ^{15}N

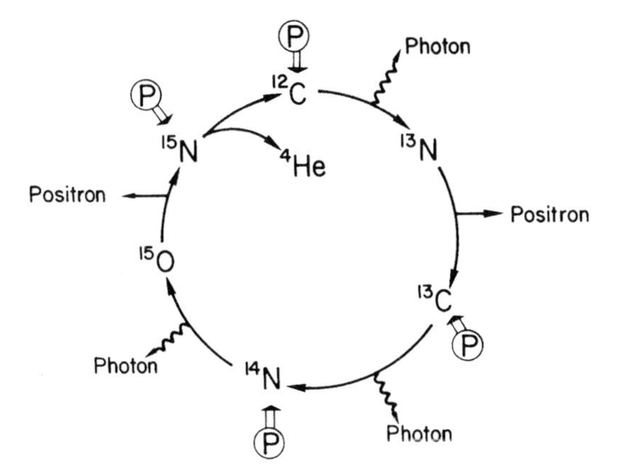

Figure 3.7 Fusion energy production in the stars by the carbon cycle.

fuses with another proton to yield a carbon-12 nucleus together with an alpha particle (the nucleus of the helium isotope ^{4}He). In this cycle, which is schematically described in figure 3.7, the main hydrogen plasma of a sun is transformed into a helium plasma while releasing energy in the form of radiation: photons, positrons, and neutrinos. Thus one can see that the plasma carbon and nitrogen in the thermonuclear cycle act as catalysts, i.e. they are not lost during the cycle process. The net result of this cycle is the formation of helium from hydrogen with the release of energy.

Although the carbon cycle is responsible for producing energy for most of the stars, there is also another sequence of reactions which can produce energy. For cooler stars, the temperature is too low to fuse carbon or nitrogen with the hydrogen plasma. This is because the protons cannot penetrate the strong Coulomb repulsions from the relatively high charge of the carbon or nitrogen nuclei. In this case, the main sequence of events is the nuclear fusion of two hydrogens to create a deuterium. The fusion processes are repeated until the end product of helium (^{4}He) is reached.

Following are a few interesting data of our present Sun and an outlook for the future:

- For about 5 billion years, our Sun has been emitting radiation (photons) with a power (energy per second) of hundreds of billion times that of the US energy consumption.
- Our Sun oscillates at all times. Sometimes the oscillation period is 5 minutes and sometimes as long as one hour. From these oscillations scientists learn about the internal structure of the Sun.
- The density from the Sun's center to about one quarter of its radius is about 100 gram per cubic centimeter. On the other hand, the outside

shell of the Sun (one quarter of the radius from its surface) is less than a tenth of a gram per cubic centimeter. The nuclear fusion is in the dense part of the Sun.
- The Sun has hydrogen fuel to burn for another 5 billion years.

What happens next? The central part will contract and produce more energy; the outer parts will expand and cool. The color of radiation will change from (today's) white to red. The Sun will become a *red giant* and later on a *planetary nebula.*

During the above expansion, the central part (that contains helium) will continue to contract, reaching a temperature of about 80 million degrees, and fusion will occur, producing carbon. More precisely, two ^{4}He fuse and ^{8}Be is obtained, which in turn fuses with another ^{4}He. The end result of all this is the production of carbon from helium. Further fusion of carbon with a ^{4}He creates oxygen. As a result, the central part (which at this stage has about half of the original total solar mass) will turn into a *white dwarf* star. The density of white dwarfs is about one ton per cubic centimeter!

Today white dwarfs can be observed in the central region of planetary nebulae through a telescope.

3.6 The Solar Exterior

The plasma interiors of the stars have appropriate conditions for thermonuclear reactions to occur naturally. Whereas the Sun's interior is extremely hot and dense, the outer layers are cooler and the density is much too low for thermonuclear reactions to take place. These outer layers represent the visible surface of the Sun and are called the photosphere (see figure 3.8). The Sun's radiation into space stems from these outer layers. A very small portion of this radiation is the sunlight reaching the planet Earth. Encircling the photosphere, to a distance of about three times the Sun's radius, there is a very hot and very low-density plasma which is called the *corona* (see figure 3.8). Corona in Latin means crown. Although the temperature of the corona is very hot, the density is far too low for thermonuclear reactions to occur. The properties of the different regions of the Sun are summarized in table 3.1.

The solar corona is visible in a dramatic and spectacular way during a total eclipse of the Sun. At this time, the Sun is hidden by the Moon in such a way that only the outer atmosphere of the Sun can be seen. A total eclipse of the Sun occurs somewhere on Earth about every year.

Although the Sun's interior is very, very hot, namely millions of degrees, the Sun's surface is only about 6000 degrees. The corona is

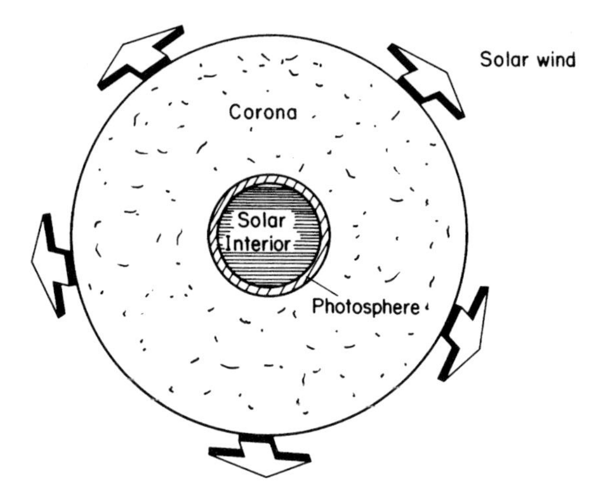

Figure 3.8 A schematic structure of the Sun: the solar interior, the photosphere, the corona and the solar wind.

observed to be in a state of motion from the Sun outwards. A 'wind' of particles, the so-called 'solar wind' (see figure 3.8), streams outward through the solar system reaching beyond the most distant planets. The corona is a completely ionized plasma of hydrogen and helium as well as small quantities of ions of heavier elements. The temperature of the corona is greater than that of the photosphere. By using today's modern instruments scientists are able to measure and to study the properties of the corona.

The solar wind is often affected by solar flares (see figure 3.9) associated with the sunspots that appear on the surface. These flares are eruptions of hot and luminous plasmas that rise to hundreds of thousands of kilometers above the Sun's surface. Sunspot activities have a cycle of about

Table 3.1 Parameters in the Sun's plasma.

	Temperature	Pressure	Density
Sun's interior	About 15 million degrees Celsius	Many billions of times larger than the Earth's atmospheric pressure	Very dense; about 150 times the density of water
Photosphere	About 6000 degrees Celsius	Low	Very low; about one-thousandth of an atmosphere density
Corona	Millions of degrees Celsius	Very low	Extremely low

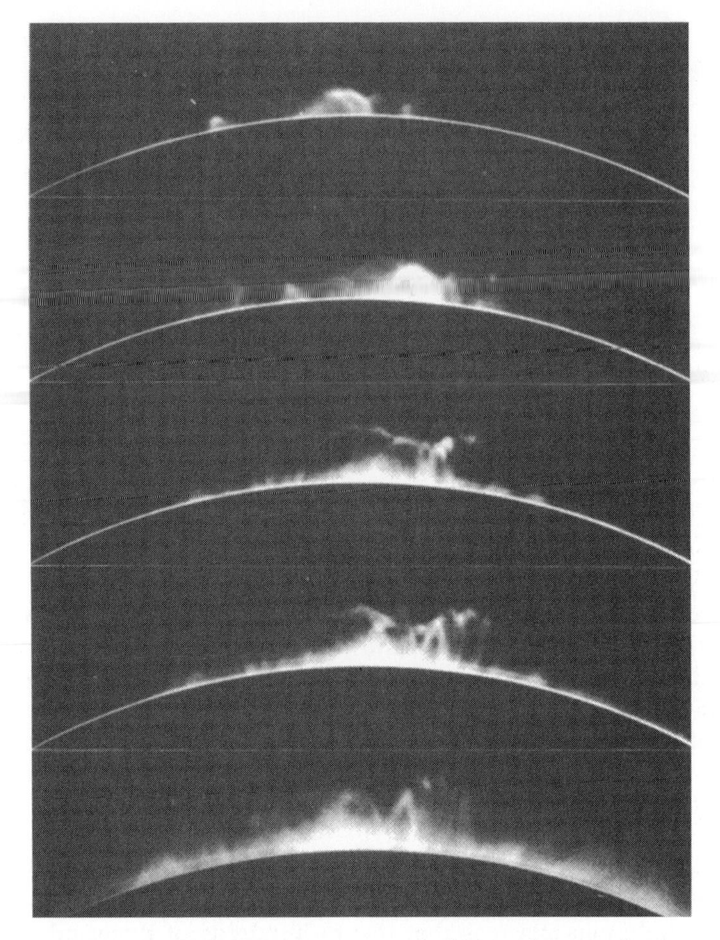

Figure 3.9 An eight-hour sequence of photos, showing luminous arches of plasma which are believed to mark the recovery of the solar magnetic field after a major flare. These loops are thought to be manifestations of a high-temperature (two million degrees Kelvin) healing process, which takes several hours to complete. In this process, the strands of the solar magnetic field which opened up during the flare gradually reconstruct and reform. Once the magnetic field has reformed, the loops fade and finally disappear. (Courtesy of National Optical Astronomy Observatories, Tucson, Arizona, USA.)

22 years. The first to observe a sunspot was the famous Italian physicist Galileo Galilei in 1610.

The magnetic fields inside the Sun's corona are considered to be responsible for the evolution of the coronal structure. Similar to the magnetic field of the Earth and many other celestial objects, the Sun's magnetic field is believed to originate in the motion of plasma charged particles in the interior of the Sun. Currents, as we already know, create a magnetic field. The flow of electric currents in the interior of the Sun causes a magnetic field in the exterior of the Sun. Every 11 years the Sun's north and south magnetic poles reverse their positions. The interaction between the coronal plasma and the magnetic fields is similar to that between plasma and magnetic forces in man-made laboratories. In both cases, electrons and ions in the plasma move in helical paths around lines of force of the magnetic fields. The charged particles move along the direction of the magnetic field lines. The Sun's atmosphere is a plasma moving in a medium of strong magnetic forces. The Sun's atmosphere (see figure 3.10) is much more violent, chaotic and complicated than our own atmosphere (which is in a gas state of matter).

An understanding of our Sun's corona is important for our understanding of the behavior of complex hot plasmas in magnetic fields elsewhere in the Universe. The development of man-made satellites has enriched the study of plasma behavior in our solar system and has helped in the understanding of the plasma physics of our Universe.

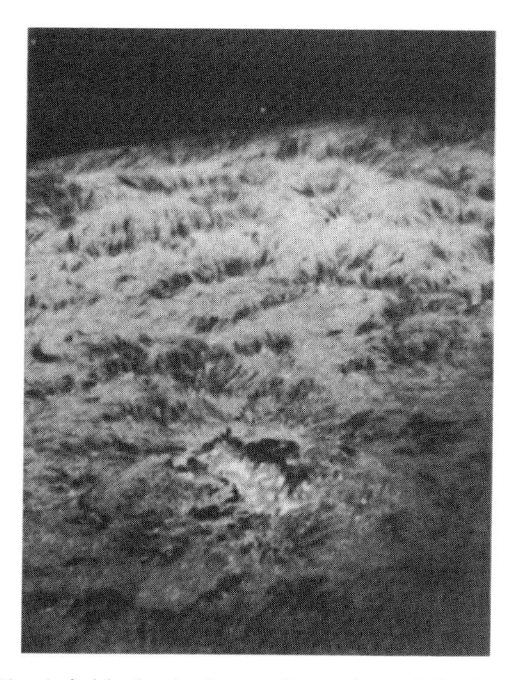

Figure 3.10 A field of spicules on the surface of the Sun in a 1971 National Solar Observatory photograph. A spicule is a short-lived (minutes), narrow jet of plasma spouting out of the solar photosphere. (Courtesy of National Optical Astronomy Observatories, Tucson, Arizona, USA.)

3.7 A Supernova Explosion

On the night of February 23, 1987, a supernova was first observed in Chile and later in other places throughout the southern hemisphere. This was the brightest exploding star seen from Earth during the past 383 years. This violent explosion occurred about 170 000 years ago in the Large Magellanic Cloud. Following the explosion, this supernova, named 1987A, was brighter than a hundred million suns. For the first time scientists could observe the most spectacular phenomenon in nature with modern and sophisticated instruments. This explosion provided new information for our understanding of the Universe.

A supernova explosion is the sudden death of a large star. The star evolves during millions of years, passing through various stages of development. During its lifetime, this star, like any other, burns through thermonuclear fusion reactions. When the nuclear fuel is exhausted, the plasma pressure is diminished and the core collapses in less than a second due to the force of gravity. During the star's lifetime, the gravitational force of attraction, whose tendency is to contract the star's dimensions, is balanced by the thermal pressure of the plasma. This thermal pressure is due to the agitation of the plasma particles, namely, the electrons and the ions, that like to expand into space. This agitation, known as the plasma temperature, derives its energy from the thermonuclear reactions inside the core of the star. When the thermonuclear reactions are reduced (due to the lack of nuclear fuel, for example) the temperature decreases, the pressure of the plasma drops, and the gravitational force causes the core of the star to collapse. The star thus collapses under its own weight. The result of this collapse is a powerful explosion, unlike any known here on Earth. It is many, many billions of times greater than any nuclear explosion (see figure 3.11).

The light from a supernova explosion is brighter than that from an entire galaxy consisting of billions of stars. When the explosion is terminated, the star is considered to have died; most of its mass is scattered into space and the left-over material is very condensed and does not shine like the other stars.

The basis for the theory of supernova explosions was introduced by the British astronomer Fred Hoyle. There are two classes of supernovae, known as type I and type II supernovae.

A supernova of type I is the explosion of a white dwarf, a very condensed star. The Indian-born American astronomer Subrahmanyan Chandrasekhar calculated that a white dwarf with a mass up to 40% greater than that of our Sun can remain stable without any explosion. If a white dwarf is a member of a binary star system then the gravitational attraction of the white dwarf can pull mass from its large companion star. When the mass accumulated by the dwarf star reaches the critical value, the Chandrasekhar mass, a catastrophic collapse of the star begins.

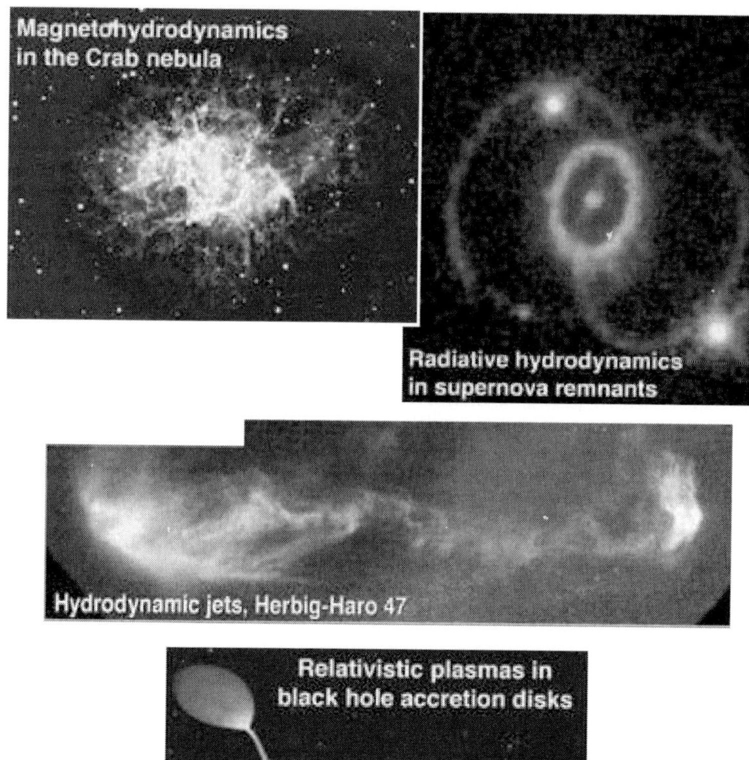

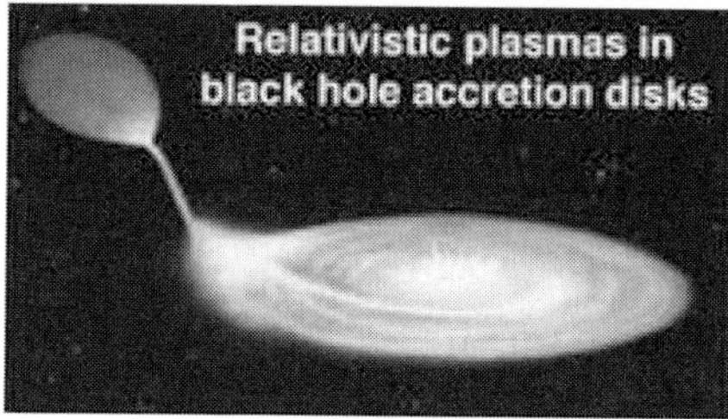

Figure 3.11 Exotic plasmas in the Universe. (Courtesy of the University of California, Lawrence Livermore National Laboratory and the US Department of Energy.)

A white dwarf is a very dense star which has exhausted its nuclear fuel and for this reason does not have thermonuclear reactions and does not shine like other stars. During the accumulation of mass from its companion, the white dwarf increases its mass until it reaches the critical mass predicted by Chandrasekhar. At this stage, the gravitational force is so large that the white dwarf starts to collapse. It decreases its dimensions and increases its density. The new higher densities are appropriate to start thermonuclear reactions in a very violent way. The energy released during this explosion is seen as a supernova.

A type II supernova is a large regular star where a series of nuclear fusion reactions take place. During the first thermonuclear reactions, four nuclei of hydrogen fuse into a single nucleus of helium. The mass of the helium is slightly less than the masses of the four hydrogens. This difference in mass is transformed into energy according to Einstein's formula. The fusion of hydrogen into helium continues until all the hydrogen is used up. The core of the star is squeezed because the force of gravitation is not in balance with the high pressure of the plasma created by the fusion energy. As a result of this compression, the core and its surroundings are heated. The core of the star becomes so hot that nuclear fusion between the helium nuclei produces carbon. Further heating of the star burns carbon to create neon, oxygen and silicon. During the last cycle of thermonuclear fusion the silicon nuclei combine to form the iron isotope ^{56}Fe which is made up of 26 protons and 30 neutrons.

Iron represents the end of the fusion processes. The fusion of ^{56}Fe and other heavier elements does not release energy. All the elements up to ^{56}Fe from Mendeleyev's Periodic Table can fuse if sufficient plasma temperatures and densities are reached. For heavier elements than ^{56}Fe in Mendeleyev's table the fusion process ceases to release energy. In a supernova, when the iron stage has been reached all the possible thermonuclear fusion reactions are finished. The plasma pressure drops and the star collapses very quickly. Elements which are heavier than iron (from iron to uranium) are also created in a supernova not through fusion but rather by nuclear reactions. In these processes neutrons (created in the supernova explosion) are absorbed and thus increase the atomic weight, while the beta decays that follow increase the atomic number.

Only the largest stars are capable of the thermonuclear set of reactions leading from hydrogen to iron. It is interesting to point out that although a larger star has more nuclear fuel it burns faster. This is because in larger stars the temperatures and plasma pressures are bigger and the nuclear reaction rate is faster. For example, the lifetime of our Sun is about 10 billion years while a star 10 times heavier will burn itself to death 1000 times faster.

A supernova of type II becomes a neutron star or a black hole following the iron stage. When the core of the star collapses a very vigorous wave in the form of a shock wave rebounds from the core of the star in an extremely large explosion. The expanding cloud can be brighter than our entire Galaxy for a short period of time. It was just such a glowing ball that was seen on our planet Earth in February 1987.

3.8 Synchrotron Radiation

How is synchrotron radiation achieved in a laboratory? Inside a circular accelerator, charged particles are confined by a magnetic field and are accelerated to high energies by an alternating electric field. While these charged particles are circulating around the magnetic field, they emit electromagnetic waves, known as synchrotron radiation.

What is the origin of synchrotron radiation in space? Space is a perfect laboratory for synchrotron radiation. The two vital elements necessary for producing synchrotron radiation, namely, charged particles inside the plasma and strong magnetic fields, are in abundance in many domains in space.

How strong are the magnetic fields that influence the plasma motion in space? The range of magnetic fields in space vary over many orders of magnitude, from extremely small to extremely large. For example the magnetic field in sun spots is about 3000 gauss (like in a vacuum cleaner motor), while in the solar wind it is about 10 million times smaller than in the sun spot. On Earth the surface magnetic field is about 0.5 gauss while in a pulsar (neutron star) the magnetic field can reach many billions of gauss.

How is the synchrotron radiation measured on Earth? This radiation, also referred to as radio astronomy, is detected by radio waves arriving from objects in the sky known as radio stars. Space synchrotron radiation also emits visible light which is measured by telescopes.

Nebula

A luminous cloud of gas and dust occurring in interstellar (between stars) space is called a nebula. These clouds containing plasma have dimensions of about ten light years. Inside the nebula a magnetic field is created by currents of electrons. Synchrotron radiation is generated by fast electrons spiraling around magnetic fields. Beautiful, visible structures of nebula due to the synchrotron radiation are detected on our planet from the crab nebula (see figure 3.11).

Cosmic Plasma Jets

The atmosphere is filled with many disturbances preventing a clear view of objects in space through standard telescopes. With the introduction in space of the NASA Hubble Space Telescope, since 1990 scientists have been able to widen the visible spectrum in order to analyze the data from the extreme ultraviolet to the far infrared. With this telescope jets (tubes of plasma) have been clearly seen. For example, in the giant elliptical galaxy M87 situated at the Virgo cluster of galaxies, a 4000 light years long jet of plasma was observed. Spiraling along the magnetic fields at velocities almost equal to the speed of light the electrons from this huge plasma jet are observed in the visible light and radio waves. The visible light and radio wave emission of these electrons are similar to those observed in a synchrotron accelerator.

One theory suggests that the above plasma jet is connected to a black hole, positioned at the center of the galaxy M87. Strong magnetic fields with a structure of a huge coil are generated in the surrounding vicinity of this black hole by a rotating and increasing disk of plasma (see figure 3.11).

According to the basic physical law of energy conservation, the amount of emitted energy is equal to the energy lost by the electrons. The energy loss by the electrons during the rotation is proportional to the 'frequency of rotation', equal to the 'frequency of synchrotron radiation', times the size of the magnetic field. The electrons spiral from the central axis outwards along the magnetic field, emitting visible and radio waves. Therefore, the electrons which produce radio emission in the plasma jet can survive much longer than the electrons which produce visible light (since the frequency of visible light is much higher than the frequency of radio waves). For the huge plasma jet in space seen in the galaxy M87, the electrons emitting visible light survive about a few hundred years before losing their energy, while the electrons emitting radio waves can survive tens of thousands of years. Therefore, the lateral size of the jet as is seen by visible light is about a few hundred light years, while by radio waves it is about tens of thousands of light years.

Other jets (see figure 3.11) with lengths of about 10 thousand light years have been observed by the Hubble space telescope as well. One interesting jet that was observed appeared to be like a twisted pair of tubes. In this case the lateral size of the jet as seen by visible light is much higher than expected. More precisely the lateral size is about 10 000 light years of visible light. This phenomenon might be explained by plasma instabilities caused by the flow of electrons along the magnetic field. For example, plasma instabilities can boost energy to the spiral electrons and thus elongate significantly the time of visible irradiation.

With the Hubble telescope in space and the radio detectors on Earth, the jets in the Universe are observed and the plasma physics in space can be analyzed on a wider spectrum.

Pulsars (Neutron Stars)

Pulsars were discovered in the 1960s by measuring radio waves arriving from the cosmos. For this discovery the British astronomer Anthony Hewish received the Nobel prize in physics in 1974.

Pulsars are stars that emit short pulses of electromagnetic radiation at regular intervals. These stars are believed to be neutron stars which rotate very quickly. While our planet makes one rotation in 24 hours, the neutron star can rotate a few hundred times in *one second*! A neutron star is extremely dense (one cubic centimeter of this star weighs hundreds of millions of tons) has a total mass of the order of our Sun and a radius of about 15 kilometers. The atmosphere surrounding the neutron star is in a plasma state possessing very strong magnetic fields. The strength of the magnetic field in different neutron stars varies between 10 million and 1000 billion gauss. The electrons in the plasma atmosphere of the neutron star rotate around the magnetic field causing them to emit synchrotron radiation. The spectrum of this radiation is very versatile, from radio waves to visible light, X-rays and even the very energetic photons called gamma rays. When the neutron star rotates, the direction of the irradiation rotates as well, enabling us to see one pulse every time the beam radiation sweeps the Earth. To quote Hans Wilhelmsson, in this way the extraordinary phenomenon of the 'light houses of the cosmos' is operating.

Two neutron stars rotating around each other at close distance emitting synchrotron radiation and other electromagnetic waves have also been observed. Due to this emission the pulsars lose energy until they collide. For instance, when one star loses energy its velocity is reduced until it 'falls' on to the second star. Such a collision of two giant nuclei produces another neutron star. In 1974 two American astronomers, Russell Mulse and Joseph Taylor, detected the double pulsar by using the radio telescope in Puerto Rico. For this discovery they received the Nobel prize for physics in 1993.

3.9 Comets

Comets are objects of our solar system which move around the Sun in very elongated, elliptical orbits. When a comet is passing not far from Earth, a spectacular moving star is seen in the sky. The comet is composed of a nucleus surrounded by gas, and has a long tail. The nucleus, with a dimension of a few kilometers, contains small frozen particles, while the luminous tail is a plasma which can be as long as a few million kilometers. By measuring the spectrum received from the emitted light of a comet, it is known that the comet contains hydrogen, nitrogen, carbon and oxygen as well as molecules of these atoms. The ionization of the comet particles is caused by protons and electrons of the solar wind.

How many comets are there in the sky? The famous astronomer Johannes Kepler suggested about 400 years ago that there are as many comets as there are fish in the ocean. Today, it is estimated that the number of comets surrounding our solar system is about 100 billion. A few comets are seen each year on Earth through telescopes. Sometimes they are seen by the naked eye.

Using the Hubble space telescope, as well as man-made satellites, great spectacles have been observed and important scientific discoveries have been made. In 1994 photos were taken by the Hubble telescope to show the comet Shoemaker–Levy 9 (named after the discoverers of the comet) which collided with the planet Jupiter. This collision caused a series of giant explosions with fireballs of more than a thousand kilometers in diameter. Geologists believe that such a comet collision on our planet long ago destroyed completely the existence of the dinosaurs.

Another famous comet, Halley, was observed during 1985–1986 by a satellite. An important observation was the Alfven plasma waves in the Halley comet tail. Moreover, the satellite instruments measured X-rays emitted from this comet. In this comet the interaction between plasma waves (in the present interstellar magnetic field) and the solar wind is accelerating electrons to the high energy necessary to create these X-rays.

It is interesting to point out that some comets contain all four states of matter—solid, liquid, gas and plasma. Such a comet was observed in 1995 by Alan Hale and Thomas Bopp and was named the Hale–Bopp comet. This comet had three tails with a length of about 3 million kilometers. One tail was composed of a plasma state of matter and emitted visible blue light, the second tail emitted white light (sunlight reflected from the dust particles of this tail) and the third tail was composed of sodium (Na) atoms. In 1997 during every night of March and April this comet was seen by the naked eye even though the nearest point from Earth was about 200 million kilometers, 50 million kilometers more than the distance between the Earth and the Sun. Comet Hale–Bopp will return to us in the year 4376!

3.10 From the Visual to the Plasma Universe

An important goal in a scientist's life is the intellectual desire to understand the Universe and the laws of nature that govern it. The scientific development of the knowledge of our Universe could be generally depicted in three phases. The first phase, which lasted for about three millennia, is the observation of the stars and the mapping of their positions in the sky (similarly to the mapping of the globe). This first phase is still being continued and developed further through the use of larger telescopes and more sophisticated technologies as they become available (like the Hubble space telescope).

The second phase, which lasted about three centuries, is the understanding of the laws governing the motion of the stars. The beginnings of this phase can be ascribed to the works bequeathed by the British genius Sir Isaac Newton. His discovery of the remarkable law of gravitation describes the force between two masses. (This force is inversely proportional to the square of the distance between the masses and directly proportional to the magnitude of the masses.) This law is universal. Using this law, together with his law for motions of bodies (that a force is equal to the mass of a moving body times its acceleration), Newton was able to describe the motions of the stars in the sky as well as the motion of objects here on Earth.

Newton's theory of gravitation was dramatically extended by Albert Einstein with his development of the theory of relativity. Einstein modified Newton's theory in accordance with his principles of relativity—time and space are not absolute. Also the force of gravitation is slightly modified in Einstein's theory. This second phase is still far from being completed. The gravitational laws related to quantum physics, the so-called 'quantum theory of gravity', are still under much investigation. Scientists are still looking for the gravitational agent, the 'graviton', which is the analogy of the electromagnetic agent, the 'photon'.

The third phase, which began in the last century, is mainly based on the fact that all the stars in the sky (except for the few planets, their moons and some other objects like the meteorites) exist in a plasma state of matter. More than 99% of the Universe is in a plasma state of matter. In 1908 the Norwegian physicist Kristian Birkeland was the first to relate experimental laboratory plasma physics to cosmic plasma physics. It is today evident that the laws of our Universe are described not only by the gravitational laws but also by the laws describing the plasma state of matter. For example, the energy sources of the stars were only understood after the development of nuclear physics and the realization that the stars are in a plasma state of matter. Our understanding of the plasma phase of our Universe is still in its infant stages. Perhaps this is the reason why most of the books about the Universe, most of the articles in our newspapers and journals, and almost all of the school text books on astrophysics choose to ignore plasma physics, although the matter under consideration is in a plasma state. Thus, since our Universe consists mostly of plasma, our understanding of astrophysical phenomena depends also on our knowledge of plasma physics.

For over 3000 years human civilization has built up its knowledge of the Universe by observing a very narrow octave of the electromagnetic spectrum, namely visible light. Infrared and radio waves were detected only during the last century and thus enlarged our view of the Universe. It is less than two decades since through space research scientists began measuring the short wavelengths of the electromagnetic spectrum: the ultraviolet, X-rays and gamma rays. The reason for this last development is that these short waves do not penetrate our atmosphere. The development of spacecraft was necessary in order to observe the short-wavelength spectrum. Today's satellites are sent into space with telescopes and modern instruments in order to acquire information about our Universe. X-rays and gamma rays are emitted from very hot plasmas in the Universe. Moreover, the magnetized plasmas cause the electrons and ions to cycle around the magnetic fields. These circular motions of the charged particles in the plasma also emit radio waves which are detected here on Earth. All these measurements of short waves in space by manned or unmanned satellites, and of radio waves here on Earth, are of radiation emitted from hot and very energetic plasma in the Universe. Even the visible light seen by the astronomers since the beginning of civilization is caused by plasmas. Visible light comes from stellar photospheres (surfaces) which are plasmas with 'low' temperatures (many thousands of degrees Celsius).

Observations of cosmic plasma by spacecraft near our planet offer an excellent opportunity for studying how plasma matter behaves in space. This study provides important information necessary to understand interstellar and intergalactic plasma phenomena. In the future, scientists will strive to achieve a better connection between laboratory plasma and cosmic plasma physics in order to unify the plasma physics research over many, many orders of magnitude: from laser-produced small 'suns' of millimeters in size to large astronomical 'suns' and galaxies. To understand and to unify physics from micro systems here on Earth to macro systems in the Universe is intellectually a very exciting and remarkable goal. The Nobel Laureate in plasma physics, Hannes Alfven, concluded his famous book on *Cosmic Plasma* with the sentence: 'The basic properties of a plasma seem to be the same everywhere, from the laboratory to the Hubble distance.' Edwin Powell Hubble was a famous astronomer who showed that the galaxies are receding from our Milky Way galaxy. Moreover, he observed that the external galaxies of our Universe have a receding velocity proportional to their distance from the Earth. The Hubble distance is used to refer to the distance between our planet and the known edge of our Universe.

Chapter 4

Plasma in Industry

4.1 Understanding Plasma for Application in Industry

Understanding plasma for application in industry has paved the way to advanced manufacturing products. Industrial plasma has not only improved the quality of life but has provided new challenges to the scientific community. By the use of the 'Fourth State of Matter' properties for applications in industry, higher quality and sometimes cheaper products can be obtained. In addition, plasma technology is environmentally cleaner than similar processes used in existing multi-trillion dollar polluting chemical industries.

A modern industrialized society exhausts large amounts of energy. Plasma-related industries could greatly increase the efficiency of energy consumption and replace some of the many existing chemical industries. Thus, by using plasma devices, large amounts of energy can be saved.

The plasma medium can tolerate a much higher temperature than any other environment. No other medium can excite atoms and molecules to radiate as efficiently nor achieve comparable non-equilibrium conditions. As a result, plasma can influence and improve some chemical and material industry. At present, the most important plasma application is in the production of large-scale micro-electronic circuits, used in almost any electronic device such as personal computers. Plasma processing of materials affects several of the largest manufacturing industries, including defense, automobiles, biomedicine, computers, hazardous wastes, aerospace and telecommunications (see figure 4.1).

How is the relevant plasma produced for industrial applications? What physical properties does this plasma possess? What are the basic processes in plasma for applications? The answers to these questions are discussed below.

(a) Direct Current (DC) Electrical Discharges

Electrical discharges were known already in the eighteenth century. By the end of the nineteenth century Thomas Edison wrote a patent in the

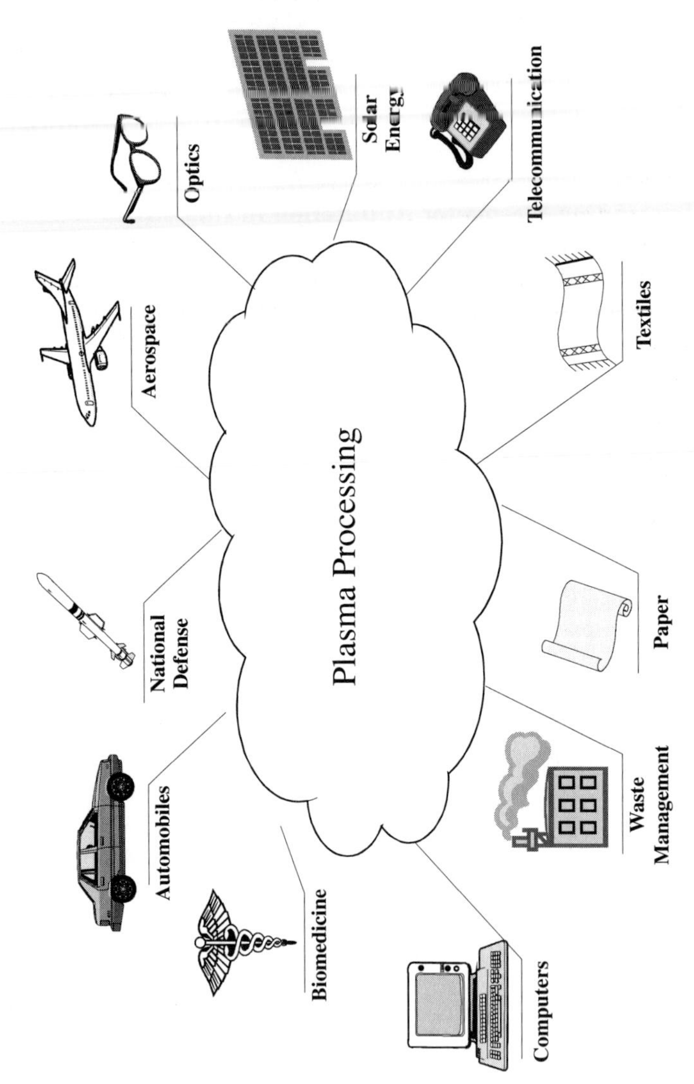

Figure 4.1 Plasma processing in industry.

USA for electrical discharge surface coating for building the first phonograph.

During the lightning in a thunder storm, very powerful uncontrolled electrical discharges occur. In order to achieve 'controlled thunder storms in a laboratory' it is necessary to begin with a vacuum-sealed tube where the density of the gas is reduced significantly by many orders of magnitude. This tube contains two metals, the anode and the cathode, connected through wires to a battery and operated by a switch. The anode is connected to the positive battery terminal while the cathode is connected to the negative one. When the switch is turned on, an electrical discharge occurs inside the sealed tube, transforming the gas (a poor electricity conductor) into a plasma state (a good electricity conductor). As a result, a DC current flows through a steady state plasma. The behavior of the plasma inside the discharge is determined by the values of the current and the potential difference (voltage) between the anode and the cathode (see figure 4.2). Listed below are a few distinctive electrical discharges:

1. The arc discharge is obtained with a high current (between 1 and 100 000 amperes) and a low voltage between anode and cathode (about 10 volts). For example, the arc discharge in a fluorescent lamp causes the excitation of the atoms in the gas to emit ultraviolet (UV)

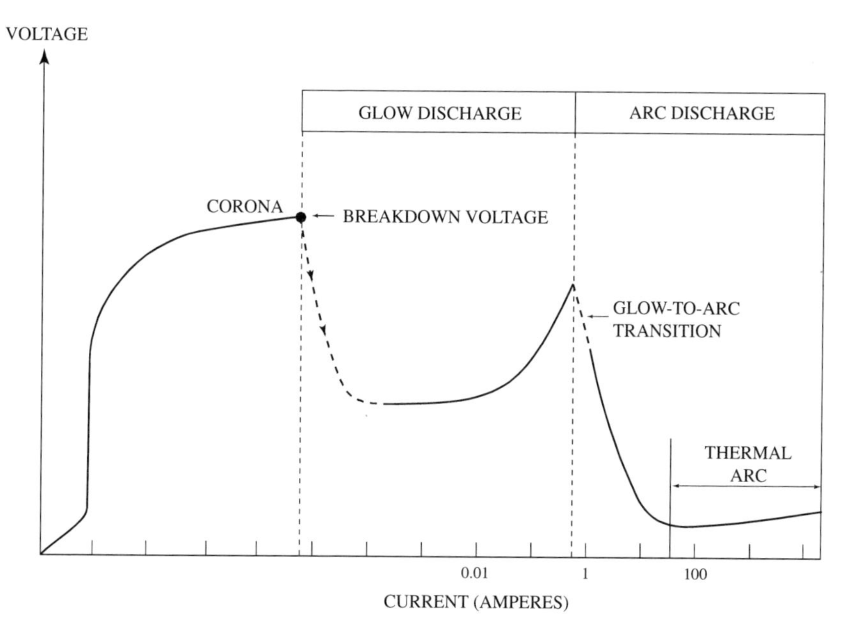

Figure 4.2 Voltage–current characteristic of a DC electrical discharge tube.

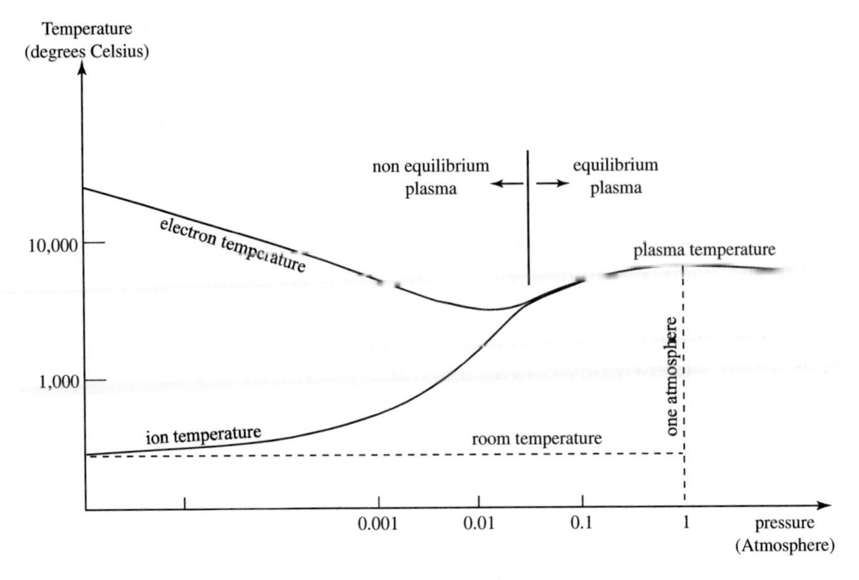

Figure 4.3 Temperature and pressure domain for equilibrium and non-equilibrium plasmas for DC discharges.

light which in turn causes the fluorescent material to irradiate visible light.

2. Glow discharge occurs at low currents (less than 1 ampere) and at high voltage between anode and cathode (a few hundred volts). For example, a glow discharge is used to clean surfaces in vacuum systems. A glow discharge is less luminous than an arc discharge.
3. Corona discharge is characterized by very low currents and very high voltages. For example, the corona discharge is the source for the blue light that can be seen in the vicinity of electrical wires of power lines.

The electrical discharges are the basis for the plasma in industry. Industrial plasmas can be divided into two categories: equilibrium and non-equilibrium plasmas (see figure 4.3). The equilibrium plasma, known also as thermal plasma, has almost equal temperatures of electrons and ions. In a non-equilibrium plasma, the electrons are hotter than the ions but the gas pressure is lower. Whereas the glow discharge is a non-equilibrium plasma, the arc discharge can be in either state.

(b) Radio Frequency (RF) Discharges

In an RF discharge the power supply interacts with the plasma almost exclusively by displacement current (the electrodes are not necessarily in contact with the plasma) instead of the real current. The radio frequency power creates displacement currents inside the plasma and delivers energy to the plasma. The RF, in comparison with the DC method, can be advantageous since the material of the electrodes may introduce impurities into the plasma.

The RF power can interact with plasmas either inductively or capacitively. The inductive case transfers the power to the plasma in the same way as a transformer transfers the power from one line to another. Schematically, a coil connected to the RF power supply is wrapped around a plasma quartz tube causing the energy to be transferred in an inductive way. This plasma is maintained in a steady state by the RF power supplied to the coil. The RF used is between frequencies of 10 kHz (kilohertz) and 30 MHz (megahertz). The gas pressure in these devices is usually below one atmosphere, although in some applications atmospheric pressure can be used.

In the case of the capacitively coupled RF plasmas, two electrodes connected to the RF power source are used instead of the coils. This method is operated with RF frequencies between 1 and 100 MHz. In this case the plasma is heated directly by the RF electric fields. Capacitively coupled RF plasma sources are used around the globe in the microelectronic industry.

One can produce a glow discharge at atmospheric pressure by using a high voltage RF at kHz frequencies. This plasma can be in a steady state at one atmospheric pressure in air or other gases and does not require a vacuum system as in the case of the DC plasma discharge.

(c) Microwave Electric Discharges

Microwaves are electromagnetic waves with frequencies (about gigahertz) much higher than the RF frequencies. High power microwave sources became available as a result of the development of radar during the Second World War. This research led to a variety of microwave sources in the order of kW (kilowatts) power operating in a steady state or in a pulse mode. In a microwave-generated plasma one can obtain a higher electronic temperature than those obtained in DC discharges or in the RF discharges. The temperatures that can be obtained within the microwave discharge are of the order of 100 000 °C in comparison with 10 000 °C in the other discharges.

The microwave-generated plasma can be derived over a wide range of gas pressure, ranging from atmospheric pressure down to very low pressures. Because the temperature is much higher in the microwave discharge plasma, the fraction of ionized gas is higher in comparison with other methods. This is an important advantage in many plasma chemical applications.

Microwave discharges have been successfully used to provide a lasing medium for efficient gas lasers such as CO_2. Gas lasers have been applied to improve high-tech industries such as optical lithography and marking. In medicine, these lasers have significantly improved surgery, eye and artery treatments.

(d) Plasma Reactors

How is the plasma produced inside a reactor? Gases are inserted into the plasma vessel (reactor). Due to an electric discharge inside the vessel, the electrons in the plasma are moving very quickly. When they collide with the molecules of the gas inside the vessel, they dissociate the molecules of the gas into atoms. Further collisions between the electrons and the atoms result in ionization, forming ions and free radicals (atoms and molecules that participate in chemical reactions). The reactions taking place in a plasma processing reactor are given in figure 4.4.

It is important to control the generation of the flux of electrons and ions inside a plasma reactor. Molecular dissociation generates free radicals

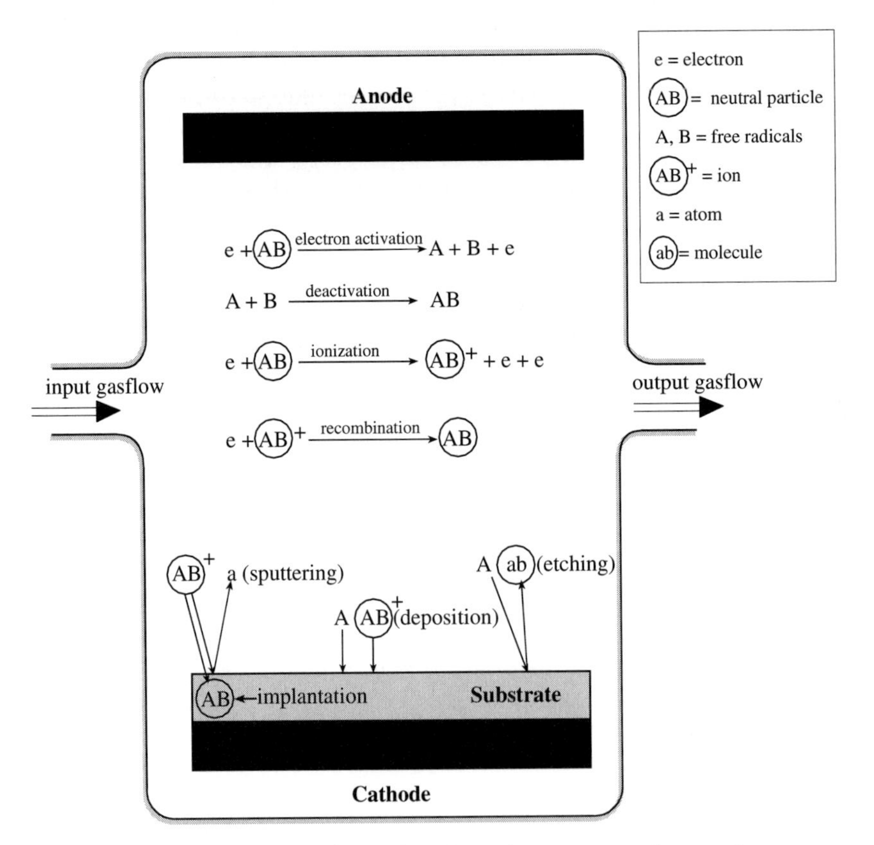

Figure 4.4 Reactions taking place in a plasma processing reactor.

that may combine to form other materials or form back into the original components. Inside the vessel, at the cathode, radicals and ions can be absorbed, bonded or sputtered. If the products of the surface reaction evaporate easily (volatile) they leave the surface and *etching* results. For involatile products, *deposition* occurs and a surface film grows. For products that remain on or just below the surface, *implantation* results.

Different reactors are designed in order to get only the desired process. For this purpose, there are many operating parameters (external variables) such as gas flow rate, gas composition, pressure inside the vessel, electrical power and frequency, magnitude and directions of external magnetic fields (sometimes used to confine or guide the charged particles inside the plasma), substrate temperature and reactor geometry. The process is determined by the internal variables, which are the result of all the external parameters. The internal variables are electron and ion densities and fluxes, electron and ion temperatures, neutral gas molecules and the free radicals produced in the reactor. The understanding and the control of the relationship between the external and internal variables are crucial in order to develop and advance this field. Due to the large number of the above parameters involved and the complexity of any plasma medium, modern computers and sophisticated numerical analysis have become crucial for the advancement of plasma-aided manufacturing.

Inside a plasma reactor, close to the cathode, a strong change in the voltage is created, inducing a strong electric field, referred to as an electrical *sheath* (see figure 4.5). (If a large change of potential occurs

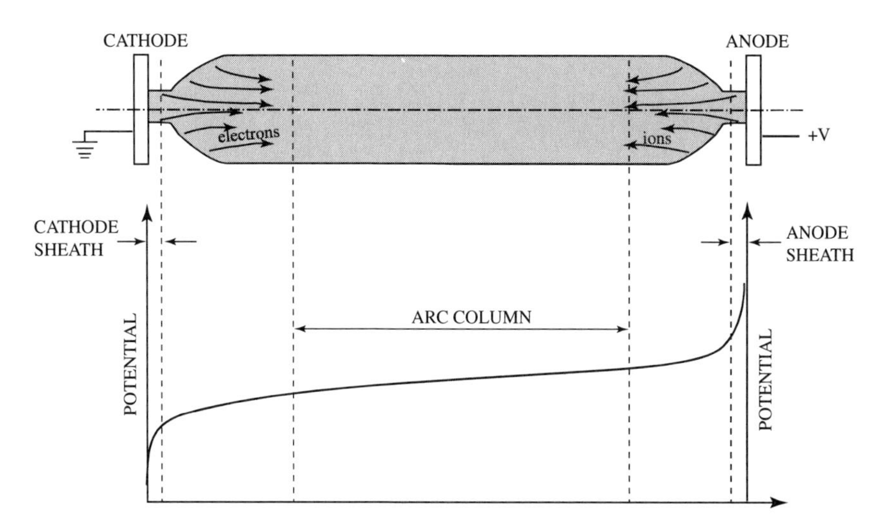

Figure 4.5 A schematic presentation of two sheath formations, near the anode and near the cathode in an arc discharge.

inside the plasma, it is called a double layer). This electric field applies a force on the ions and accelerates them towards the cathode. As the accelerated ions strike the cathode, atoms are ejected from the cathode towards the plasma. These sputtered atoms are then deposited on the surface of a substrate material, producing a thin film.

The rate of thin film formation on the substrate depends on the amount of sputtering of the cathode, which in turn depends on the potential drop in the sheath. It is imperative that the plasma components do not chemically react with the target or the substrate. For this purpose inert gases are used as the plasma medium. (Inert gases such as He, Ne, Ar, Kr, etc. appear in the last column of the Mendeleyev table.)

The unparalleled properties of the Fourth State of Matter such as high temperatures, energetic electrons and ions, and the existence of plasmas in a very wide range of densities, pressures, and temperatures often provide the only practical use in industry. Present plasma applications are discussed:

Semiconductor electronics
Plasma modification of materials
Plasma spray
Plasma welding, cutting and material processing
Plasma space propulsion
Plasma display panels
Plasma and the diamond industry
Treating waste with plasma
Plasma lighting

Future possible applications are also discussed:

Particle accelerators
X-ray lasers
Isotope separation
Plasma antennas

4.2 Semiconductor Electronics

In 1944 the first semiconductor transistor was discovered by John Bardeen, William Shockley and Walter Brattain who received the Nobel prize in 1956 for this discovery. In July 1948, the last item of 'The News of Radio' column in the *New York Times* reported that a transistor was demonstrated in a radio receiver which replaced the conventional tubes for the first time. Many years passed before this half-inch long transistor was miniaturized to a few micron size (a micron is one millionth of a meter).

Today, the semiconductor electronics industry consists of the production of almost all of the high technology electronic devices such as computers, televisions, radios, cellular phones, etc.

What is a semiconductor? A semiconductor material is not as good a conductor of electricity as copper, yet not as bad as an electrical isolator such as gas. Why then has it become so essential for the modern high-tech industries? It appears that this 'not so good conductor' can be manipulated easily by doping it with impurities (elements from the third or fifth column in the Mendeleyev table of elements). These few impurities change drastically the electrical properties of the material, transforming it into a good conductor. By using two different semiconductors having different impurities, the system can act as a rectifier or as a good amplifier of small currents. Furthermore, in a similar way other electronic devices are produced.

The typical semiconductor material is silicon found in the sand on the beaches. The famous Silicon Valley industry in California is the result of the high-tech revolution caused by semiconductors.

For semiconductor electronics, plasma-aided manufacturing has become essential. This plasma industry has a multi-billion dollar market per year. For this industry very large integrated circuits (chips) are produced. An integrated circuit is one tiny device capable of many electronic functions. In the manufacturing of this integrated circuit the plasma ('dry') processes replace the chemical ('wet') processes.

Plasma etching processing for microelectronics has become a very important technique for producing very tiny semiconductor devices. Wet etching is achieved by using acids in a liquid state of matter. Dry etching is achieved by using plasma. In the latter case the etching is faster, more accurate and can be done in any desired direction (anisotropic). In plasma manufacturing, more electronic components such as transistors and diodes can be placed on a microchip. In order to achieve smaller, faster and cheaper chips, anisotropic etching is crucial. Dry processing is required for smaller separation between circuit elements. Since a critical electronic device is of the order of one micron, a dry processing technique such as plasma etching is suitable for this purpose. Whereas standard etching uses chemical acids, plasma etching is a cleaner environmental process.

The widest use of plasma in industry today is the production of microelectronic devices.

4.3 Plasma Modification of Materials

Plasma modification of materials is used to improve the surface properties without changing their bulk properties. These processes are carried out by ion implantation into the material. This method has become economically alluring in high tech centers and can be accomplished on metals and alloys, semiconductors, ceramics, insulators and polymers.

Plasma-Assisted Chemical Vapor Deposition (PACVD)

The refractory (i.e. unmanageable, difficult to melt, etc.) metals are best deposited by plasma using sputtering or *plasma-assisted chemical vapor deposition* (PACVD). Plasma is used in a reactor to initiate chemical reactions in a gas with an electric discharge. For a standard chemical vapor deposition (CVD) the interaction between the substrate and the vapor occurs at high temperatures. Sometimes these high temperatures damage the substrate. By using the PACVD method the ions are attracted from the (hot) plasma toward the substrate, which can be kept at much cooler temperatures than in the CVD method.

In a plasma atmosphere it is possible to produce materials that are difficult or impossible to produce with conventional chemistry. Furthermore, while in a pure chemical industry there are many subsequent processing steps, such as synthesizing, deposition, etc., in a plasma reactor all these processes are accomplished simultaneously in one step.

Plasma Polymerization (PP)

PP is similar to PACVD. The main difference is that in the PP case the deposition is for organic materials (polymers) while the PACVD is for inorganic materials. For example, when tetrafluoroethylene gas is injected into a plasma, a Teflon-like film is deposited on a substrate. The plasma turns this starting gas into fragments which either nucleate into a polymer film at the surface of the substrate or polymerize into a chain of clusters.

Under suitable conditions, monomers (units of one compound) introduced into a plasma environment can combine to create polymers (many repeated units of one or more compounds). Plasma processing can produce thin films of polymers necessary in many important applications such as coating on other materials, multiple layers for magnetic recording tapes or disks and plastic wrapping materials. Here too the magnetic recording industry is a multi-billion dollar enterprise.

The coating with the above polymers is very important also in the biomedical area, making it possible for the development of new prosthetic devices, implants and tools for medical monitoring and diagnostics.

Ion Implantation

Surface properties of materials are improved without having to change their bulk properties through implantation of specific ions into the material. The following surface properties can be modified by plasma-assisted chemical vapor deposition:

1. the hardness of the material
2. fatigue (extreme weariness from prolonged exertion or stress)
3. toughness (the ability to withstand great strain without breaking)
4. adhesion (state of being united)
5. friction (the resistance of a surface to its relative motion with another surface)
6. dielectric properties (insulator materials capable of maintaining an electric field with minimum loss of power)
7. corrosion (deterioration through chemical action)
8. resistivity (the ability of opposing an electrical current)
9. oxidation (the ability to combine with oxygen)

A plasma source can be used to extract ions in order to implant them into the materials to be modified. A strong electric field exists in a plasma near the cathode (the cathode sheath); see figure 4.5. This electric field accelerates the positive ions arriving from the plasma into the cathode. If inside the plasma tube the pressure is kept very low (in a low density plasma), the ions can be accelerated to strike the cathode with very high energies of 100 keV (kiloelectronvolts) or more. At this level of energies the ions can penetrate the cathode and be implanted inside. In this way, the hardness and lubricity of tools are improved. In a similar way, at lower energies, this technique is used for the micro-electronic industry.

It has been shown that the ion implantation can modify and improve the mechanical properties of steel. Fatigue life of tools can be extended by as much as a factor of two. The friction of tools can be reduced by up to a factor of 100 and therefore the wear resistance increased by a significant amount. Most implantation is done with nitrogen ions; however, the use of carbon, boron, titanium and molybdenum have also shown good results.

Today, scientists believe that it is cheaper and better to modify the surface of a material than to change its bulk properties. This can be better achieved by plasma modification of materials.

4.4 Plasma Spray

Plasma spraying is used to apply a thick coating to substances that do not yield to other treatments. With the aid of plasma spraying an effective anti-corrosion coating is achieved, especially for high temperature industrial applications. Losses due to material corrosion alone can amount to more than 50 billion dollars per year in the United States.

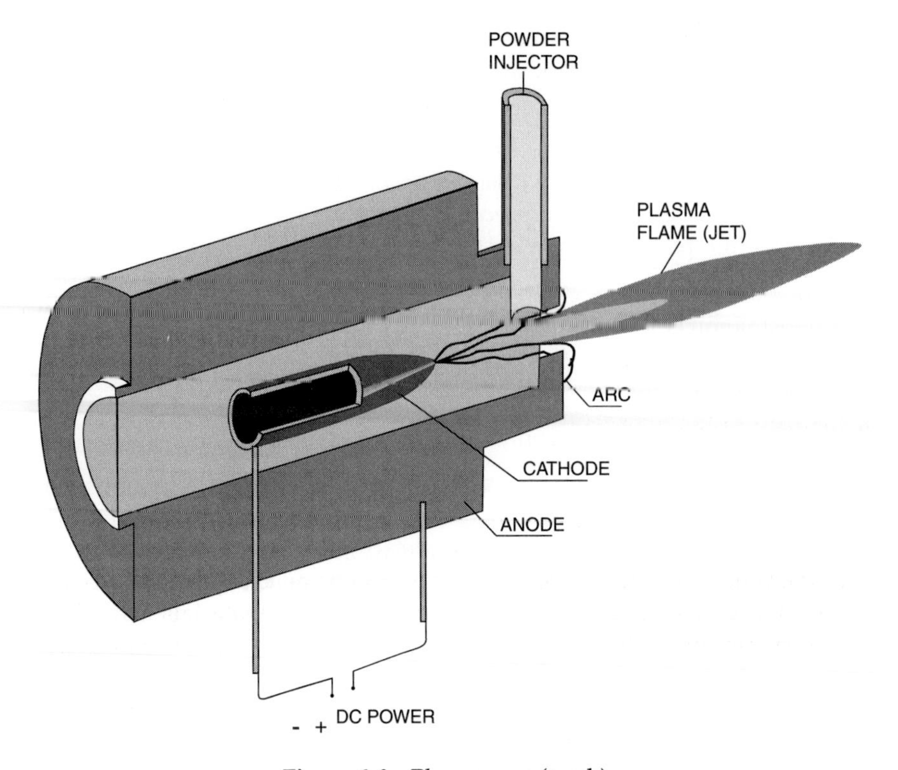

Figure 4.6 Plasma gun (torch).

Plasma spraying is done with a plasma gun, also called a plasma torch (see figure 4.6). This technique provides surface protection to turbine blades in aircraft engines, as well as many other tools and materials. The material is protected against corrosion by superalloy spraying and thermally insulated with ceramic spraying.

The plasma gun consists of two electrodes, the cathode having a cone shape and a cylindrical anode which extends beyond the cathode to form a nozzle (see figure 4.6). An inert gas is blown into the gun between the electrodes which is then ionized to form a plasma stream (jet) through the nozzle. The gun is operated when a current creates an arc between the electrodes. A DC power supply of about 50 volts sustains a steady current which is hundreds of amperes.

The coating material in powder form is injected into the plasma jet. The hot plasma flame melts the powder particles and accelerates them toward the target to be coated. These melted droplets hit the target surface, where they solidify and accumulate to form the desired protective coating. It is necessary to keep the plasma at the right temperature in order to prevent the powder particles from overheating, which can cause them to evaporate. Due to electron and ion collisions an equilibrium state is obtained, reaching a temperature of about 15 000 °C.

For successful application of the plasma gun, it is necessary to control many variables such as the size of the powder particles, the temperature, the pressure of the carrier gas, the geometry of the injection of the powder, etc.

After the powder is heated and melted in a plasma gun, it is accelerated by a plasma jet toward a substrate. On impact the powder forms a coating. Almost any material that can be melted without becoming decomposed can be applied. Plasma spray coating of ceramics and even polymers can be done in air, in vacuum and even under water. In short, plasma sprayed coatings are used in internal combustion engines, in power plants, in industrial machinery, in aircraft engines, and in general any area where extreme demands on materials are required.

Besides plasma spraying, other methods of coating such as electrolysis (galvanizing) or vapor deposition are also applied. In the latter case the coating material is vaporized and condensed on to the substrate atom by atom. Whereas standard procedures cannot be used with all materials, plasma spraying can be applied to a large variety of materials including refractory ceramics. In comparison with other high temperature procedures, the plasma technique is safer for coating, since the plasma gas is chemically inert (does not interact and deteriorate the material) and the target can be kept fairly cool. Technologically, the plasma gun is not significantly more sophisticated than a paint sprayer.

Plasma gun applications are used to coat materials for better thermal insulation. For example, a quarter of a millimeter layer thickness of ceramic material can protect the engine of an aircraft from overheating. Different parts (piston caps for example) of large diesel trucks, boats, and locomotives have been plasma sprayed with ceramics in order to allow the engines to operate at higher temperatures and thus increase their efficiency.

Another application of plasma spraying is to produce a thin coating of metal to protect turbine blades from oxidation, which leads also to rusting. By coating two layers of different materials one can get better thermal isolation as well as avoid oxidation. For example, on many engines a thin layer of chromium or aluminum is added to a ceramic layer. The ceramic layer serves as the thermal isolation while the other layer protects from oxidation.

One of the most frequently used applications of the RF produced plasma is the inductive plasma torch. Compared with the DC plasma gun, the inductive plasma torch is capable of treating larger surface areas with lower gas velocities. The lack of direct contact between the plasma and the heating coils in the inductive plasma torches, in contrast to the DC plasma gun, enables cleaner plasma processes without contamination of the hot gases in the arc jet. This method is applied to high purity material production, high temperature thermal treatment of material, surface coating, chemical vapor depositions (CVD) at atmospheric pressure, etc.

Although plasma spraying appears to be a simple process, the science of spraying is far from being simple and far from being completely understood.

4.5 Plasma Welding, Cutting and Material Processing

Thermal plasma devices are widely used in welding and cutting hard materials. These operations can be carried out in the factory and on site. In welding, both the substrate and the filler metals are melted in order to achieve a strong metallurgical bond. In order for this method to be successful, both materials should have similar melting points. The welding process is also used for thick overlay (mm) coatings. A list of weld surfacing materials is given below:

Low alloy steels (used for agriculture equipment)
High speed steels (for cutting tools)
Stainless steel (in chemical plants)
High chromium steels (for brick and cement machinery)
Nickel alloys (for hot work tools, valves, chemical plants)

Although welding with plasma torches gives much stronger and more durable results than coating deposited by other techniques such as flame spraying, some problems exist. For example, inclusions from the electrodes can cause porosity (full of pores).

High pressure plasma in an equilibrium state between electrons and ions is used for basic material processing in industries such as melting, evaporation, and dissociation of minerals (such as oxides, carbides, and sulfides). The above processes require high temperatures for some materials that can be achieved only with plasma. High power plasma guns operating at temperatures of up to 15 000 °C have been developed to melt any known material.

There are several advantages to plasma material processing. Thermal plasmas can operate at high pressure even above one atmosphere, thus the loss of rapidly evaporating materials can be avoided. As a precise control of power can be attained in a plasma device, this leads to the control of exact temperatures yielding accurate processing of materials. Since the energy density in a plasma can be higher than in other states of matter, faster melting is achieved. The plasma can be an inert or chemically active gas medium.

Plasma heating is also useful in the refining of steel where accurate control of atmosphere, pressure and temperature is very effective in order to reduce the excess impurities such as phosphorus and sulfur.

Technological demands have led to the improvement of plasma welding, cutting, and material processing.

4.6 Plasma Space Propulsion

The propulsion of space vehicles, like that of any other rocket or jet aircraft, is based on momentum conservation. Momentum is equal to the product of mass in kilograms (mass is the weight of the object as measured on earth) and its velocity in meters per second. When a vehicle moves in space and releases gases, or any other mass, with velocity in one direction, this lost momentum causes the vehicle to move in the opposite direction. The momentum thrown out of the space vehicle is the momentum gain by this vehicle. According to the famous Newton Law of Motion, this momentum divided by its time of action is the force acting on the vehicle and is usually called the thrust. Thus the larger the mass or the velocity of the jet stream coming out of the vehicle the larger the propulsion achieved.

Plasma space propulsion has been studied for more than three decades. The principle is based on ejecting electrically heated plasmas out of the vehicle. This type of plasma propulsion requires less fuel mass than existing chemical systems, potentially making space exploration less expensive. As plasma can be heated to very high temperatures the jet velocity can be increased, thus decreasing the weight of the fuel required.

Another propulsion device, whereby ions are accelerated out of the vehicles, has worked successfully for more than 15 years for maintaining orbit satellites in space. For this purpose, low power (of the order of 100 watts, like a bulb) thrusters are necessary to overcome small friction losses to keep the track on the trajectory. For interplanetary missions, significantly higher power thrusters will be required. The problem so far is to develop electrodes with lifetimes longer than exist today for longer space travel.

It was suggested to use the plasma outflow from the Sun to fly through deep space. This would be similar to a sailor using sails for maneuvering in the winds on the high seas. There is a strong wind of plasma blowing through the solar system, namely, hot ions and high energy particles are blasted outward from the storms in the Sun. This is called the solar wind and it whistles past the earth at more than 500 kilometers per second. Researchers suggest using this wind to 'sail' through space. About 30 years ago NASA engineers studied ways to build kilometer size sails using thin plastic sheets coated with aluminum.

Plasma space propulsion is a promising option for interplanetary missions. Scientists in this field are confident that in the future plasma space propulsion will be better, cheaper and faster.

4.7 Plasma Display Panels

The demand for flat panel televisions has led to successful research for plasma display panels. For small, compact panel displays, the liquid crystal has been used successfully for some years. However, in order to manufacture large size panels of this kind without defects, using liquid crystal is technologically difficult and very expensive. Moreover the light emitted from a liquid crystal panel is limited. Therefore the picture can only be viewed at a very narrow angle perpendicular to the screen. A person sitting on the side, not facing the screen, will not see the picture clearly. In order to overcome the above obstacles, scientists have developed plasma display panels. The depth and weight of a 40 inch plasma display would be 5 cm and 10 kilograms as compared with 100 cm and 150 kilograms for the standard 40 inch television.

Today the dominant TV display technology is based on cathode ray tubes. In this tube electrons are accelerated, hit a fluorescent screen and their energy is converted into visible light. This visible light appearing on the TV panel is composed of many (about 500×500) pixels (the smallest element of an image that can be individually displayed on a computer or television screen).

In a plasma display panel the light of each picture element is emitted from a plasma created by an electric glow discharge. The voltage pulse between the electrodes leads to the breakdown of the gas and a formation of a weakly ionized plasma. As a consequence, this plasma emits visible and ultraviolet light. A weakly ionized plasma represents a complex system of electrons, positive ions, excited atoms of the gas and photons which interact among themselves and with the induced electric field inside the plasma. The ultraviolet light emission from the discharge is used to excite a phosphorus material to irradiate in the three fundamental television colors (red, blue, and green). As in standard televisions, the mixture of the three colors is the same for the plasma display panels.

Although plasma flat panels are produced today, they are still very expensive. To make them economically available, further understanding and development of the relevant weakly ionized plasma system is required.

4.8 Plasma and the Diamond Industry

For hundreds of years, diamonds have been 'a girl's best friend'. Today, diamonds have become 'a very good friend of industry'. While women prefer diamonds to be big and clear, the diamonds used in industry are in powder form or thin layers.

What is a diamond composed of? Diamonds are made of pure carbon, geometrically arranged in a tetrahedral form. Diamonds are the hardest substance existing in nature, good heat conductors (the thermal conduction of diamonds is about five times higher than copper), electric isolators, optically transparent and almost inert against chemical interactions.

Because of their hardness, diamonds have many important industrial uses, crucial to the metal working industry. Following are some of the benefits derived from the use of diamonds in industry:

1. Their two properties, good heat conduction and bad electronic conduction (electric isolators) are used for the microelectronic industry. In the 'chip' industry, it is important to isolate electrically the different components and to get rid of the local heat.
2. The optical transparency of diamonds along a wide spectrum range, from infrared through the visible light up to the ultraviolet spectrum, offers multiple applications in the optical industry.
3. Due to their hardness and their lack of chemical interaction, diamonds are used as coating for blades, mechanical tools, etc.

What role does plasma play in the diamond industry? By using plasma deposition processes, the development of diamond coatings was made possible. A plasma discharge in a gas of carbon atoms, for plasma-assisted chemical vapor deposition, is used to produce diamond layers. The plasma medium is created from atomic hydrogen and carbon. The graphite carbon or a cheap gas containing carbon such as CH_4 (methane) are used. This carbon is deposited on a substrate above the cathode. In the plasma reactor the hydrogen and the carbon are ionized by collisions with the electrons of the plasma. Although the positive carbon and hydrogen ions are then deposited on the substrate, the hydrogen atoms 'evaporate' while the carbon remains in the shape of a diamond.

The end product of diamonds, a 'good friend of industry', is put to use in the high-tech industry. It is therefore not a surprise that diamond powder and diamond layers have become a multi-billion dollar market.

4.9 Plasma and Treating Wastes

A major problem facing mankind around the globe is the disposal of hazardous waste materials. Many billions of tons of wastes exist around the world, in anticipation of better and cheaper means of 'clean-up'. Plasma can be used to 'clean up' some of the waste materials categorized below:

1. Concentrated liquid organic hazardous wastes. These include some chemicals used for processing paint solvents and cleaning agents.

2. Solid wastes contaminated with organic hazardous materials. In this class of wastes one can find contaminated soils, medical wastes, containers filled with hazardous liquids, such as capacitors and transformer fillings, warfare agents, etc. Furthermore, heavy metal contamination requires special treatment to prevent the pollution from reaching the ground water.
3. Wastes from manufacturing processes. In this category, very disturbing wastes are the metallic dusts acquired from the metallurgical industry. These wastes are extremely dangerous because the dust of materials such as lead causes air pollution and ground water contamination.
4. Low level radioactive or mixed wastes. These wastes must be deposited in solid containers that are non-leachable for very long periods of time. If the containers are not sealed off perfectly, radioactive gases such as cesium escape into the atmosphere.
5. Municipal solid wastes. These enormous quantities of pile-ups are not only contaminated with heavy metals but they also contain combustible materials.
6. Hazardous chemicals in gas media. For example, some of the gases are destroying the ozone layer in the atmosphere.

Following are plasma processes for the disposal of some of the waste problems listed above.

Plasma processing can be used to break chemical bonds and decompose hazardous wastes. This is presently accomplished by passing the waste material through an arc plasma either in the form of a liquid spray or a fine powder. Because plasma can be heated to the highest temperatures the undesired wastes can be decomposed into non-hazardous components.

By using hot plasma to melt the solid hazardous wastes, a stable glass is achieved that can be safely stored or even used as a construction material.

Large plasma arc furnaces have been developed using plasma torches in the treatment of hazardous wastes. Moreover, on a smaller scale, plasmas generated by electron beams have been used selectively to destroy toxic wastes such as the hazardous carbon tetrachloride molecules. By bombarding this material with electrons from the generated plasma, the molecules break up into less stable compounds that further break down into non-toxic materials.

4.10 Plasma Lighting

As soon as man realized that natural light was not sufficient for his needs, the search for artificial light began. From the bonfire, the candle light, the kerosene lamp, to the development of electricity, man has come a long way. Today lighting has even reached inside the human body (by using fibers) for diagnosis and surgery.

Before the end of the nineteenth century physicists, and in particular Sir William Crookes in England, were searching for visible light by experimenting with DC electrical discharges in sealed vacuum tubes. The physicist Georges Claude in Paris inserted neon gas inside the tube and received a red light during the electrical discharge. Neon signs soon appeared on the exteriors of the commercial buildings in the cities of the world. This was the first commercial use of plasma lighting.

Experimenting with mercury vapor in electrical discharges emitted a blue light. These discharge tubes had little application until the 1930s, when fluorescent tubes were developed: the quantity of light was multiplied immensely. From this moment on, the plasma (fluorescent) became practical for illumination of interiors of homes. How was this plasma achieved? The plasma in the fluorescent lamps is at a very low density and emits ultraviolet (UV) radiation. When the UV hits the fluorescent materials coating the inside of the lamp, visible light is emitted.

Other light sources, such as mercury lamps, are plasmas at high densities. In these devices, the plasma is irradiating directly visible light. The conversion of energy into light is more efficient in the high density plasma devices than in the low density plasma lamps. However, development of other lighting sources is needed in order to avoid the use of undesired polluting materials such as mercury.

More recently high intensity discharges and lower power compact fluorescent lamps have been developed as light sources. Some of the light sources are microwave discharges or RF discharges without using electrodes.

Inside a television tube electrons are accelerated between the electrodes. When the accelerated electrons hit the fluorescent screen they excite atoms causing them to irradiate when returning to their equilibrium state.

Radiation from lamps can have important applications in environmental clean-up such as water purification. Plasma light lamps emitting UV cause the decomposition of micro-organisms or polluted organic materials in the water. In this way the above polluting wastes are converted into water and gases (such as CO_2) that are evaporated in the air and the water becomes purified. It is important to note that efficient, cheap sources of lighting can accelerate the food growth.

At the beginning low temperature plasma was developed as a fringe benefit from the research done in the field of plasma lighting. Historically plasma lighting has been the catalysis for much of plasma research in industry.

4.11 Particle Accelerators and Plasma

In 1932, the American scientist Ernest Orlando Lawrence build a circular compact accelerator (cyclotron) for accelerating protons to energies up to 1.2 million electronvolts. He was able to disintegrate some of the nuclei of the atom and to produce artificially new isotopes. For these achievements he was awarded the 1939 Nobel prize in physics. Lawrence's cyclotron, about 25 cm in circumference, was continuously developed for newer models. Today's gigantic 27 km circumference accelerator at CERN in Geneva, Switzerland, is used for particle physics research in collaboration with European countries.

Another American physicist, one of Lawrence's students, Luis Walter Alvarez, winner of the 1968 Nobel prize in physics, contributed significantly to the development of linear accelerators and discovered many short living isotopes.

Particle accelerators are devices that are used to accelerate charged particles such as electrons, protons, and ions to very high energies. The accelerator is the basic instrument for experimental nuclear and elementary particle physics. The properties of the elementary forces in nature and the large variety of elementary particles were detected by colliding particles at high energies.

Today's accelerators are also used in other branches of physics and other sciences in order to study the properties of the solid state of matter, material engineering, etc. Accelerators are also used in medicine for producing radioactive isotopes for the diagnosis and treatment of tumors. They are also used in the food industry for sterilization, and in metallurgy for the detection of defects.

In order to accelerate charged particles in a plasma, a specific wave is required. This wave must have an electric field in the direction of the moving charged particles. Moreover, this wave must move with the same speed as the charged particles (this is similar to surfing on a wave), usually almost equal to the speed of light. This is achieved by two lasers which dictate the dominant plasma wave suitable for this acceleration. The plasma wave frequency achieved here is much lower than a laser frequency. When combining two high-frequency lasers, the result is a wave oscillating with a frequency which is equal to the difference of the two laser frequencies ('beat wave'). The difference of the laser frequencies should be equal to the plasma wave frequency; this is the proper condition for accelerating charged particles as described above. In this way an electric field can be achieved that will accelerate an electron beam in a very compact way.

At present particle accelerators via a plasma medium are not available. Today, the physical principles and the technology of these accelerators are being investigated. When these accelerators are developed, their size will be very small in comparison with today's existing accelerators. For example, it is expected that the dimensions of existing accelerators of many kilometers will be reduced to the size of meters or less, getting closer in size to Lawrence's original accelerator. Moreover, energies which are unattainable today will be available with plasma accelerators.

4.12 X-Ray Lasers

In order to have a better understanding of the molecular structure of all forms of life, including man himself, it is essential to 'see' and examine the individual molecules in living tissues. For this purpose, X-rays are needed.

Today most people have had the opportunity of observing a hologram, a three-dimensional picture reconstructed by laser light. The size that one can see with a laser is of the order of the laser wavelength, so that the shorter the wavelength of the laser, the smaller the specific structure that can be examined. In order to study biology, life cells, and genetics, a laser with shorter wavelengths than exist today is required.

To observe the structure of molecules one needs lasers of wavelengths comparable with the size of atoms. Moreover, very short durations of laser light are needed in order to study motion. Very brief X-ray laser pulses could 'freeze' molecular motion, similar to what we obtain with the 'pause' switch on a video recorder, which holds the picture for examination. In this way one could learn about the thermal vibrations of solids and the motion inside materials. Thus, X-ray lasers could revolutionize many subjects of biology, chemistry, physics, and material science.

It is extremely difficult to obtain the conditions necessary for X-ray lasers. The first difficulty is the fact that matter is opaque to X-rays, so that the X-ray photons would be absorbed in the media that 'house' the laser. Thus X-ray photons are lost inside the medium and cannot come out as in the usual lasers (optical or infrared). However, if a plasma is created inside the medium, the X-ray photons can be coaxed out. The second difficulty is the very high pumping energy required to obtain a population inversion for the X-ray laser. Population inversion is when there are more electrons in the upper state of an atom than in the lower state. Pumping energy means getting the medium ready for the laser action by such means as different electrical discharges. For example, the population inversion of a glass laser is created by flash lamps whereas in a gas laser this can be obtained by electron beams and microwave discharges. The population inversion should be between inner shells of the atoms of the heavy elements of Mendeleyev's table, where the energies between atomic levels are equal to the high energies of the photons in an X-ray laser beam. Energetic photons ionize the material and a plasma is created.

The medium for an X-ray laser is a hot plasma. This hot plasma is created by very powerful existing lasers with an infrared or visible wavelength. Atoms possessing many electrons are ionized and a population inversion is established in this plasma. For example, a high-power laser can ionize many of the 34 electrons of a selenium atom. In particular, when ionizing 24 electrons and exciting the other 10 electrons a population inversion was established with selenium. After the powerful laser creating the plasma had stopped its irradiance, the plasma suddenly cooled down. During the plasma formation or the cooling-down process (due to collisions between the atoms), a population inversion was established and X-ray lasing was observed. The novelty of this laser scheme is that the medium is a plasma, whereas the media for existing lasers are solid, liquid, or gases.

4.13 Plasma Isotope Separation

Isotopes are atoms possessing the same numbers of electrons and protons but different numbers of neutrons. In other words, these are elements with identical chemical properties but different weights, therefore one generally cannot use chemical methods to separate isotopes. Appropriate forces which depend on the isotope masses must be applied.

The huge electromagnetic mass spectrometer (Calutron) was developed during the Second World War as part of the Manhattan Project. The isotope of uranium-235 was separated from uranium-238 to obtain the 'fuel' for the atomic bomb. In such a spectrometer ions having different masses are deflected in a magnetic field into different orbits and the isotopes are separated and collected in different vessels. This simple process is not very efficient and the production rate is very small. More efficient processes were invented after the war.

There are isotope separations in use for non-nuclear materials besides the nuclear materials. For example, different isotopes are needed for medical diagnostics, chemistry, and basic scientific research. Furthermore, the demand for isotopes for medical application is increasing constantly.

Research on rotating plasmas started in 1958 at the Royal Institute of Technology in Stockholm, Sweden. In 1971 the Swedish physicist Bjorn Bonnevier reported experimental evidence of isotope separation in a rotating plasma. A few years later other groups of scientists around the world managed to achieve higher and higher efficiencies in separating isotopes by rotating plasma.

Lasers are also used to ionize the isotope, forming a low-temperature plasma. The separation process involves selective ionization of one isotope and collecting it for practical use.

Today, isotopes for medical application are still not achieved by plasma isotope separation. For the future, plasma isotope separation may become more efficient and contribute to the 'medicine of tomorrow'.

4.14 Plasma Antennas

In the *New Scientist* of November 1999, it was reported that soldiers could radio without detection thanks to a plasma antenna. Picture the following scene. You are inside enemy territory and you must radio your base. One of the biggest causes of being detected by the enemy is your long range radio antenna. If you are equipped with a plasma antenna you can signal your message without being detected. For this purpose the antenna metal is replaced by a plasma confined inside a tube similar to a large fluorescent light.

How is the plasma obtained? This is accomplished by radio wave transmission into a dielectric tube containing a noble gas such as argon via an electrode of the base. The radio waves ionize the plasma inside the tube, stripping electrons from the gas molecules, thus ionizing them to form a plasma. During transmission, the electrons of the plasma oscillate exactly like the electrons in the standard antennas so transmission of the radio signal is possible. Similarly it is possible to tune the antenna to receive incoming radio waves.

Why is it difficult to detect this plasma antenna? When the antenna is turned off, the plasma inside the tube instantly becomes a neutral gas, similar to air, and therefore cannot be detected by radar. The radar detects metals and not gases. While radar can detect metal radio antennas at all times, the plasma antenna may be detected only during transmission or reception.

4.15 More Efficient, Unique, More Environmentally Clean

In order to keep up with today's high technology plasma can contribute significantly. Plasma possesses some unique characteristics relevant for industrial application. The highest temperatures can be obtained only through the plasma state of matter, therefore thermodynamic forces (relevant to all chemical reactions) can be made more efficient. Energy densities in plasma surpass any other technique created by chemical means. These characteristics speed up the chemical processes. New energetic species are produced through plasma sources: energetic ions that can change and influence chemical reactions, a variety of photons of different wavelengths, free radicals which are ions that are active in chemical reactions, highly reactive neutral atoms (such as oxygen and fluor), molecular fragments and energetic electrons.

The above properties are used to achieve faster, and sometimes cheaper and better, products using plasma in industrial engineering. In some instances, plasma can achieve new processes not found in any other method. Furthermore, these are usually accomplished without the undesired pollution and toxic wastes.

More efficient energy utilization can be achieved in plasma chemistry, material processing with thermal plasma and in plasma lighting devices. **Unique** results can be obtained in new materials from plasma chemistry, plasma etching and deposition for micro-electronics, material processing with thermal plasma, surface modification of material, and in space propulsion systems. **More environmentally clean** processes can be obtained in plasma chemistry, plasma etching and deposition for micro-electronics, material processing, and ion implantation.

Chapter 5

The Solution to the Energy Problem

5.1 Soylent Green

If man is to survive, he needs energy. Energy provides us with electricity, heat, transportation, communication, and a multitude of other things. However, most important is the fact that energy enables us to acquire food. Food is grown in developed countries on fertilized soil using sophisticated machinery developed through advanced technology. Modern agriculture is sometimes defined as a 'process of transforming energy into food'. In order to feed the vast world population of about five billion people of today and to provide for the growing number of the future, an enormous amount of energy is required. Today, we can obtain energy by burning coal and petroleum on a large scale and from the Sun, waterfalls, or windmills on a small scale. In developed countries, energy is also acquired from nuclear reactors by the burning of nuclear fuel such as uranium. But what will happen when most of the energy supplies of today become exhausted? Without an alternative solution, civilization will be destroyed (see figure 5.1).

This chilling fact was well portrayed in the movie *Soylent Green.* The year is 2022. New York is extremely overpopulated. The city feeds its hungry, poor, and homeless on government-rationed synthetics and plankton derivatives made by the Soylent Company. Because there are few sources of energy, the city of New York is in darkness. Electricity is heavily rationed and most of the average homes are lit by a single light bulb. Very often this limited power supply is switched off entirely by the electric company and to avoid total darkness the people are forced to pump in electricity manually. Without electricity, the modern luxuries to which we are accustomed today have almost vanished. No trees or parks or gardens are anywhere in sight. The trees were cut down for heating homes during the winter. The parks and gardens were destroyed to provide room for the dilapidated housing. Most people live like rats in run-down structures. Food is very expensive. It is very seldom if ever that people can afford to buy fruit and vegetables. History is returning to the Dark Ages.

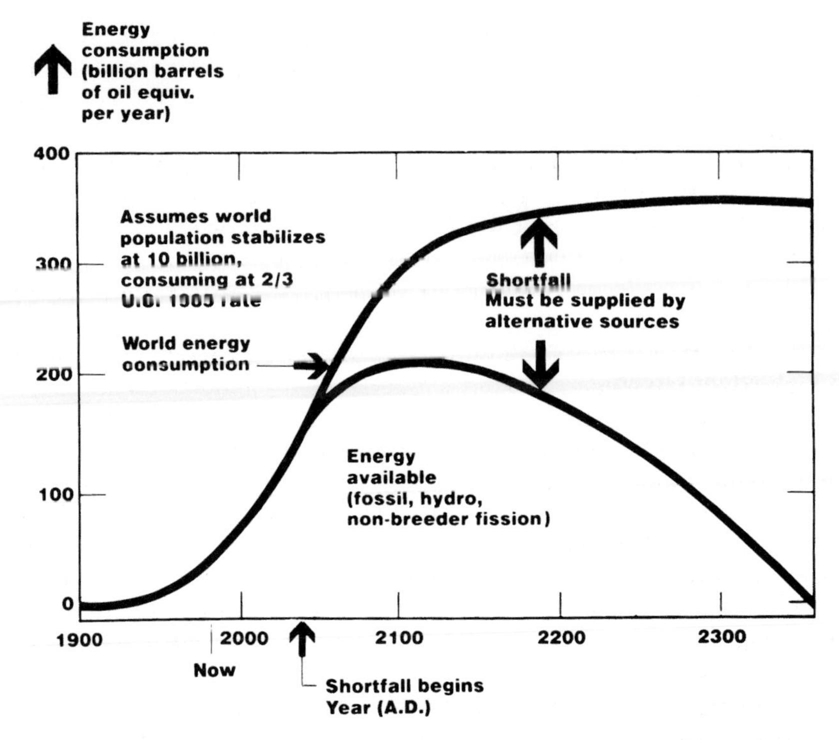

Figure 5.1 The energy consumption of our civilization. This projection shows that the fossil fuel era is almost over. If we continue to burn fossil fuels for energy, they will last only another few hundred years. At our present rate of use, experts predict a shortfall in less than 50 years. (Courtesy of Lawrence Livermore National Laboratory, California, USA.)

One of the favored and rich employees of the Soylent Company who still enjoys some of the rare modern luxuries is Simonson. When he is murdered, the investigation is taken on by a tough detective named Thorn and his researcher, Sol Roth. Roth discovers something very suspicious about Soylent's methods of producing food. The Soylent Company owns a special clinic enticing people to come to die peacefully. They are simply put to sleep (poisoned) pleasantly, while listening to their favorite music and watching a movie depicting the beautiful and splendid green scenery that existed before the energy was exhausted. After they die, their bodies are processed through special machines which then produce a synthetic food substitute which is used to feed the people. Although this movie is pure science fiction, it might, unfortunately, turn into reality if we fail to resolve the energy crisis of the future.

Many of the energy resources of our planet Earth have already been used up in large quantities and there are estimates that about half of the oil reserves have already been burnt. It now appears that in the near future we will burn the second half of our oil reserves. In the last century oil has been responsible for our high standard of living and for being able to feed the Earth's growing population (see figure 5.1).

The uncertain state of our future energy resources has prompted the scientists of the advanced nations to search for alternative sources of energy. These alternatives must be capable of supplying energy in a reliable way without disturbing the environment. This supply should also be available for a long time. Nuclear fusion has the potential to fulfil these requirements.

Scientists believe that the large amounts of energy needed to sustain our growing population can be achieved through nuclear energy. In order to understand this crucial and most significant statement, let's outline the supply requirements for a single 1000 megawatt electrical power plant (standard plant) for one year (see figure 5.2). A coal-fired electrical power station uses about two and a half million tons of coal a year which have to be transported by 25 000 railcars. Imagine the energy necessary just for the transport of this coal from one country to another. An oil-fired electrical power station uses about 11 million barrels of oil which have to be shipped by 11 supertankers. The supply of oil and coal involves similar volumes of fuel transportation. By contrast, nuclear energy reactors require much less fuel. A uranium fission plant requires only one railcar load of about 30 tons of uranium dioxide per year. The supply of fuel for a fusion plant can be carried in a pickup truck!

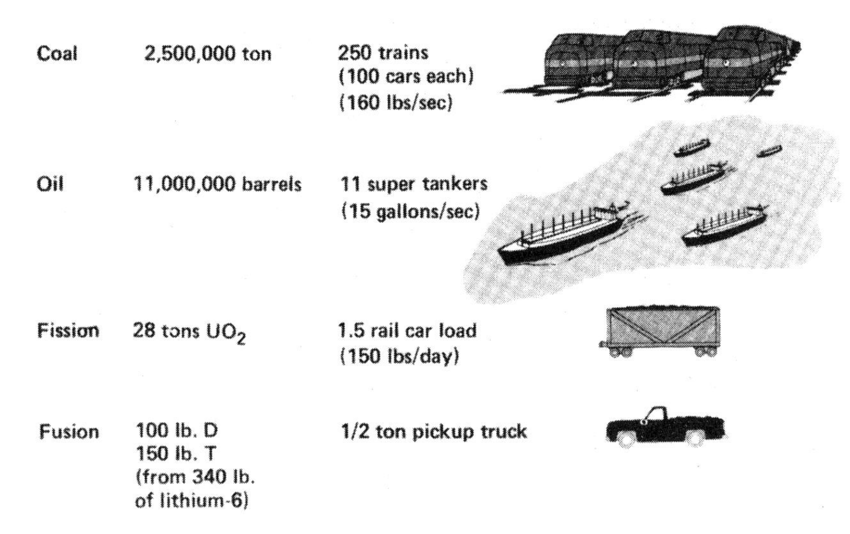

Figure 5.2 Annual fuel requirements for a 1000 MW_e power plant. (Courtesy of Lawrence Livermore National Laboratory, California, USA.)

Moreover, the environmental damage from a coal or an oil plant is very disturbing. The pollution from these plants may also change the climate, causing irreversible damage. One coal-fired power plant produces 2000 railcars of ash per year and about 100 tons of poisonous sulfur oxides are absorbed into our atmosphere every day from this one power plant alone. Oil plants cause similar atmospheric pollution. There are no comparable 'ashes' and chemical pollution from either fission or fusion plants. However, whereas fission nuclear power stations have the problem of radioactive waste, fusion plants are clean reactors, both chemically and radioactively.

5.2 World Energy Consumption

World energy (power × time) consumption per capita, defined as the average annual total power consumption in a country, divided by the number of its inhabitants, was estimated in 1995 as follows:

1. Canada and Norway with a consumption of about 13 kW (kilowatts).
2. United States with a consumption of 11.2 kW.
3. Japan with a consumption of 5.7 kW.
4. Europe with a consumption of 5 kW.

China is using about 1 kW while India is using about one third of the consumption of China. Impoverished countries utilize as little as 100 watts. In 1995 the world average power consumption was 2.1 kW. In order to know the total amount of power world consumption, it is necessary to multiply 2.1 kW times 6 billion people, yielding a global total power of about 1.2 terawatt (one tera equals a million millions). It is estimated that in 50 years from now at least three times as much power consumption would be required.

Where does all the energy consumed today come from? About 90% of the present energy consumption is derived from burning fossil fuels. This could be a serious problem in the future for two main reasons:

1. The fossil fuel reserves are limited. Without increasing the rate of consumption, the crude oil will last for about 40 to 50 years, natural gas 60 to 70 years and coal 270 years. Furthermore, one has to remember that the increase in the factor of two in the consumption rate will decrease these numbers by the same amount.
2. The most dreaded problem is that these fossil fuels are releasing gigantic quantities of CO_2 into our atmosphere, thereby causing a 'greenhouse effect' (the heating up of our planet). In one year more than 30 billion tons of CO_2 are released into the atmosphere by burning fossil fuels.

Since the beginning of the Industrial Revolution, the atmosphere has been polluted and the CO_2 has increased by 25%. In 1796, the famous Scottish inventor, James Watt, perfected his steam engine which found wide use in manufacturing. This invention was a crucial step in the Industrial Revolution in the second half of the eighteenth century.

What is the danger of the increased CO_2 in the atmosphere? As the visible light of the Sun heats the ground on Earth, the rays are converted into infrared and reflected back into the atmosphere. This causes an equilibrium between the energy coming from the Sun and the energy reflected from the Earth, yielding the present average temperature around the globe. An increase in the CO_2 gas in the atmosphere will absorb the infrared irradiation reflected from the earth causing this temperature to rise.

What happens when the average temperature increases? The answer is far from simple, but scientists fear that there is a danger that glaciers on the poles will melt and infuse fresh water into the oceans. This could disrupt the ocean currents and long Siberian winters could replace the today's existing pleasant winters in western Europe. Furthermore, raising the sea levels could cause flooding around the world. Increase in temperatures could induce more desert formation and as a result less food producing area, leading to hunger and poverty. This could very well lead to international insecurity and provoke major migrations of population from one area to another.

The continuous burning of fossil fuels causing global climatic changes might be very dangerous and irreversible to our only irreplaceable atmosphere.

5.3 Nuclear Energy

During Albert Einstein's studies in 1905 on the theory of relativity, he came to the conclusion that mass and energy are equivalent. Einstein's theory expresses the basic principle that mass may be converted into energy and energy may be converted into mass. The conversion factor between mass and energy is huge, so that a small amount of mass produces a very large amount of energy. For example, if one could convert a spoonful of sugar (about a mass of one gram) entirely into energy it would produce the same energy as one obtains from burning about 10 million liters of gasoline! From Einstein's theory one concludes that energy can be liberated in nuclear processes, whereby the sum of the masses of the products after the reaction is less than the sum of the masses of the interacting particles before the reaction. By measuring the masses of the atomic nuclei, it is deduced that there are two different ways to decrease the nuclear mass and to convert it into energy.

One way is to split a heavy nucleus into two nuclei of intermediate mass—fission; the second way is to fuse two light nuclei into a heavier one—fusion. The fission reaction uses neutrons to split a heavy atom, such as uranium, into smaller atoms. During this process, there is a release of energy and the appearance of more neutrons that can further split other nuclei of uranium (see figure 5.4). (This is referred to as a chain reaction.) This process of nuclear fission, which was discovered in 1939, revealed six years later the high concentration of energy in the explosion of the atomic bomb. The nuclear reactor developed in 1942 by Enrico Fermi at the University of Chicago in the USA was constructed even before the atomic explosion. Energy was liberated as heat and converted into electric power in a controlled way. Special precautions must be taken, however, in the operation of a fission power plant. The undesirable effects of radioactivity during the operation of the reactor and the leakage of radioactive waste must be prevented completely. The development of nuclear reactors had a troubled career and the public questions the reliability of fission nuclear technology. This was aggravated by the disastrous accident at the Chernobyl reactor in the Soviet Union in April 1986. In this nuclear accident, the control system in the reactor ceased to function and the safety back-up procedures were faulty. As a result, large quantities of radioactive waste escaped past the core of the reactor into the atmosphere. The more modern reactors in the Soviet Union, as well as all the reactors in the West, are well protected by an external 'containment' around the reactor chamber which is supposed to prevent the radiation material from leaking outside. A similar accident occurred in the USA in 1979, the Three-Mile Island accident. However, little radiation leaked out owing to the external containment structure around this reactor. The population was therefore not affected by the radiation as in the Chernobyl case.

Today there are over 500 nuclear power stations spread out over the globe. The nuclear fission energy program will remain with mankind and contribute to the electricity production system, at least until a new and more reliable energy system can be developed. Scientists believe today that such an alternative can be based on nuclear fusion energy.

5.4 Nuclear Fusion Energy

All atoms are constructed of three basic particles: the proton, the neutron, and the electron. The hydrogen (H) atom is composed of one proton and one electron only. However, the hydrogen has a 'family' of atoms which includes two other types of hydrogen: deuterium (D), which has one proton and one neutron in its nucleus, and tritium (T) with one proton

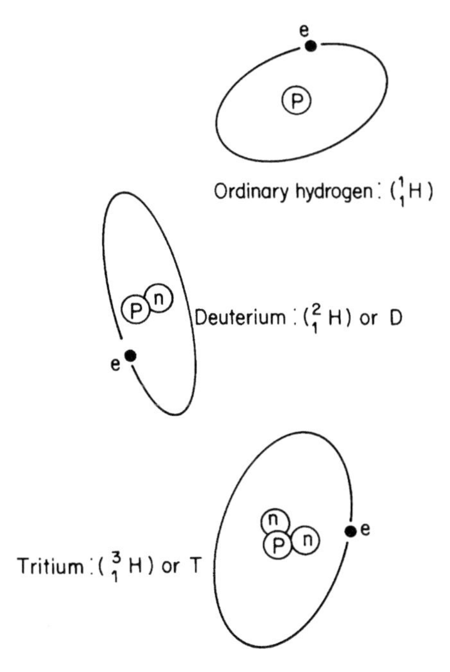

Figure 5.3 The isotopes of hydrogen.

and two neutrons in its nucleus (see figure 5.3). The chemical properties determined by the single outer electrons of these atoms are identical. However, their nuclear interactions are not alike.

Deuterium occurs naturally in water; one can extract about a pound of deuterium from some 30 000 pounds of water. Thus the enormous volumes of our oceans contain huge quantities of deuterium. Calculations show that one glass of ocean water contains enough deuterium for nuclear fusion equivalent to the energy obtained from 300 liters of gasoline. The efficiency of the release of energy by the fusion of deuterium is very high. About 50 liters of fused deuterium (obtained from about 1500 tons of ocean water) could supply all the electricity requirements of the United States for one hour. Therefore, this raw material can provide an unlimited source of energy. But what is nuclear fusion? How can one achieve it?

The basic principle of the fusion process is opposite to that of fission; instead of splitting the nucleus, one fuses two nuclei. The key ingredients for fusion are the very light elements, in particular the hydrogen family. The synthesis of two light elements to form a heavier nucleus can result in the release of large amounts of nuclear energy. Nuclear fusion processes are responsible for the colossal energies generated in our Sun and in the other stars of our Universe.

We know that inside the nucleus dwell protons and neutrons which are generally referred to as nucleons (nucleon may be considered the 'surname' of both the proton and the neutron). These nucleons are bound inside the nucleus. We learned in Chapter 2 that the Coulomb force causes protons to repel one another while the neutrons are not affected by this electric force. As the protons are positively charged and the neutrons are neutral, why does the nucleus stick together and not explode? Looking at the rhyming verse in the appendix (p. 176) on the four forces, one recalls that the strong interaction (force) is responsible for keeping the nucleus bound. The total mass of these bound nucleons is slightly less than the total mass of the same constituents outside the nucleus. (Whereas inside the nucleus the nucleons are 'jailed', outside they are 'free'.) In order to convert mass into energy, Einstein concluded that it has to be multiplied by the square of the speed of light. Thus, according to Einstein's formula, the above mass difference is equivalent to the binding energy:

$$\text{Binding Energy} = \text{Mass Difference} \times (\text{Speed of Light}) \text{ squared.}$$

The binding energy differs for various nuclei. The binding energy per nucleon is calculated by taking the above binding energy and dividing it by the number of nucleons inside the nucleus. The energy released per nucleon in the fusion process is usually larger than that in fission. Energy can be gained either by fusing the light elements or by fission of the heavy elements. The nuclei of the elements in the middle of Mendeleyev's Periodic Table are most stable and thus possess the greatest binding energy per nucleon. Iron-56 is the most stable nucleus. For example, by fusing deuterium and tritium (see figure 5.4) one gets a more stable nucleus, helium-4. Before this interaction (D + T) there are five nucleons (two protons and three neutrons); the same number of protons and neutrons exist after the fusion process (^{4}He + n). Nonetheless, the energy released is due to the fact that the binding energy of D and that of T are together less than that of ^{4}He (n in the fusion product is a single particle possessing no binding energy). We thus see that the energy released during the fusion process (as well as in the fission process) is due to the difference in the binding energies per nucleon in the various nuclei (see figure 5.5).

In the fusion process, the nuclei are positively charged and therefore they repel one another. In order for these two repelling nuclei to collide, they must be made to move with high velocity relative to each other to overcome the electric barrier. The repelling force (Coulomb force) becomes stronger for the nuclei possessing higher charges. For example, the hydrogen family contains one proton in each nucleus and therefore has a charge of one unit, the helium nucleus has two protons and thus its charge is double, or two units. Lithium has three protons inside its nucleus and therefore has a charge of three units. The Coulomb repulsive

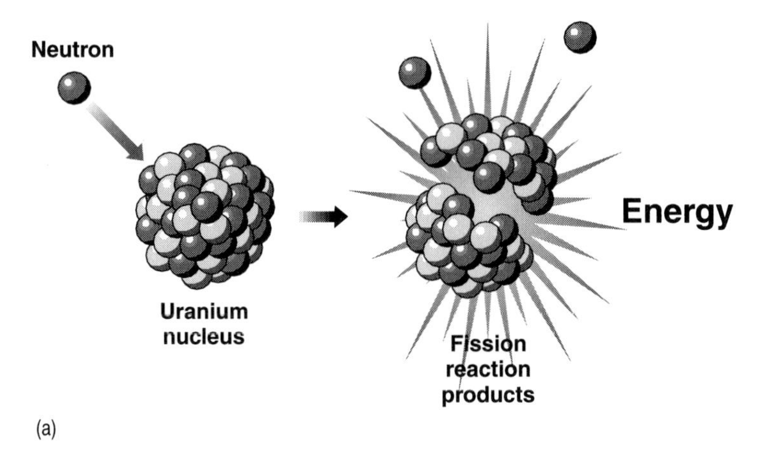

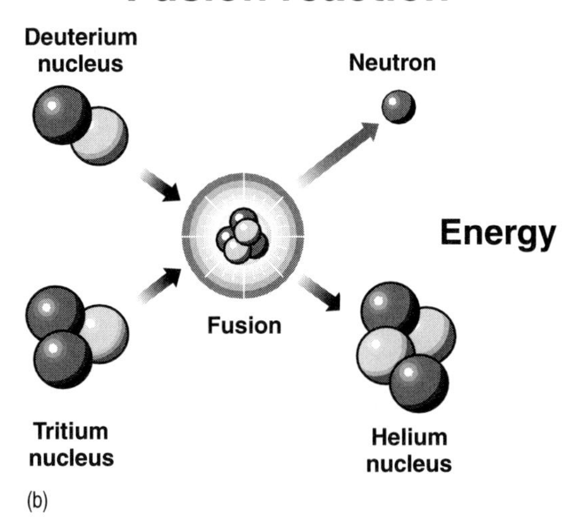

Figure 5.4 (a) Fission of a uranium nucleus; (b) fusion of deuterium–tritium nuclei. (Courtesy of the University of California, Lawrence Livermore National Laboratory and the US Department of Energy.)

forces are proportional to the charges of the interacting nuclei and it is thus harder to fuse helium than it is hydrogen. Thus nuclei containing more protons possess higher positive charges and the fusion process becomes more difficult owing to the large Coulomb repulsion forces. It is for this reason that the hydrogen family is the best candidate for

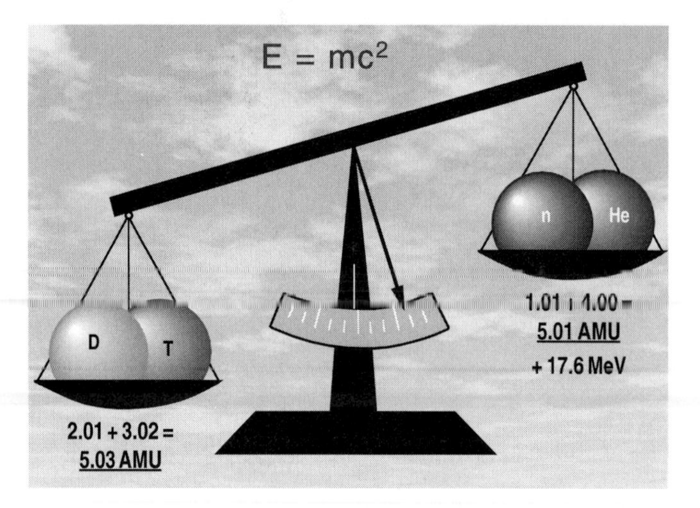

Figure 5.5 The mass of fusion fuel before the reaction is more than what is left after the reaction. The difference has been converted from mass to energy, following Einstein's famous equation (AMU = atomic mass units). (Courtesy of the University of California, Lawrence Livermore National Laboratory and the US Department of Energy.)

fusion in the laboratory. Fusion of deuterium and tritium takes place faster than the fusion of other combinations, such as hydrogen–hydrogen, deuterium–deuterium, etc.

Because of the low cost and the availability of deuterium, it would be preferable to use this isotope alone. Moreover, tritium is not stable and therefore it does not exist naturally. It is produced artificially in the laboratory (see figure 5.6).

The fusion of two deuterium nuclei can form a helium-3 (^{3}He) nucleus and a neutron (denoted by n) or they can fuse into a tritium (T) and a hydrogen nucleus, the proton (p). The tritium formed in this way can fuse very fast to create a helium-4 (^{4}He) nucleus (consisting of two protons and two neutrons) and one neutron. In the above processes there is a very effective release of energy.

In the Sun and other stars, hydrogen is the nuclear fuel. The Sun possesses extremely large quantities of hydrogen and only a tiny percentage of this hydrogen is fused. The total energy released is still immense. However, our 'man-made sun' in the laboratory or in a fusion reactor is very small and thus requires a high percentage of its nuclear fuel to burn very fast in order to release significant quantities of energy. Hydrogen is thus not a good candidate for achieving nuclear fusion energy. It is known that the D–T fusion has the fastest rate and it is the 'easiest' to achieve in a controlled way. However, as mentioned previously, tritium

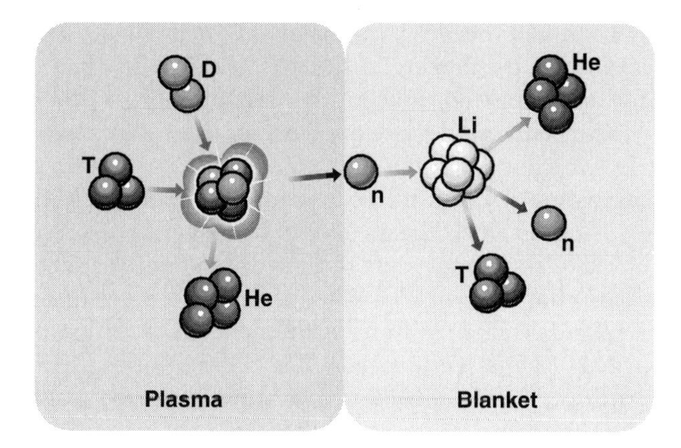

Figure 5.6 (a) The deuterium–tritium fusion; (b) the neutron from the DT fusion is used to create the tritium element, not found in nature. (Courtesy of the University of California, Lawrence Livermore National Laboratory and the US Department of Energy.)

is not available from natural sources and therefore one has to produce it in the laboratory in an artificial way. This is accomplished by the interaction between lithium (Li) and a neutron which is released in the D–T fusion (see figure 5.6). The raw materials for D–T fusion are deuterium and lithium. Large amounts of lithium are present in the oceans and on Earth. There is enough lithium for almost one thousand years of energy supply using the above process. Furthermore, scientists are working on D–D fusion, as well as other processes, which should be realized before the lithium supply is exhausted.

The end product of the nuclear fusion of the hydrogen family is the harmless helium gas and neutrons. The neutrons can collide with other materials, thus causing them to become radioactive.

Radioactivity is the undesired side effect of nuclear energy. Radioactive elements are produced either by splitting an atom or by irradiating it by neutrons. The splitting of heavy atoms is by far the more dangerous case. In a fission reactor plant, the radioactivity is produced by splitting, for example, uranium atoms. There is no way to control the undesired products caused by this process. For example, one of the fission products is strontium-90, which is absorbed by human bones and tissues. This isotope has a very long lifetime and emits gamma rays which destroy cells in the body and cause great and irreversible damage. These isotopes remain inside the reactor and produce nuclear waste. Security precautions are taken so that this waste does not leak into the outside atmosphere. In a fusion process radioactive material is not produced directly but can be produced indirectly by the neutrons. There are materials that do not become radioactive when struck by neutrons. These materials can be chosen in the construction of the core so that the neutrons are captured within the reactor core without causing radioactivity. Consequently, there are none of the radioactive fuel waste disposal problems that are common to fission reactor power plants.

A very interesting process is the fusion of deuterium with helium-3 (^{3}He). This fusion is very clean, since it does not involve neutrons which might make some materials radioactive. Moreover, both products of this fusion are charged particles (the nuclei of hydrogen and ^{4}He). Thus it is convenient and efficient to transfer their energy directly into electrical energy. However, ^{3}He is very scarce. The ^{3}He required for this process has to be produced artificially. For example, it can be extracted from the fusion of deuterium.

The fusion of deuterium–deuterium (D–D) requires a temperature higher than that of deuterium–tritium (D–T), while deuterium–helium-3 (D–^{3}He) requires a higher temperature than the D–D fusion. The higher the temperature the more difficult it is to achieve and control the thermonuclear fusion. An even higher temperature is needed (two billion degrees) for the fusion of hydrogen and boron-11 (^{11}B). Boron-11 is the common isotope (80.4% of all boron) found in seawater in huge quantities; it is only eight times scarcer than deuterium. Boron–hydrogen fusion reactions do not yield neutrons and the energy is carried by the charged particles. The energy of the charged particles can be converted into electrical energy very efficiently. Hydrogen–boron fusion would be the ideal solution to the energy problem if it could ever be achieved in a terrestrial laboratory.

In short, controlled thermonuclear fusion can supply human civilization with an unlimited source of energy. The abundance of the basic raw material for these processes, deuterium, is effectively unlimited. The oceans possess enough deuterium for more than a thousand million years of energy supply. Moreover, the cost of the raw material is negligible. A very important feature of controlled fusion reactors is the safety of these devices. There is no fear of a nuclear accident, as in the case of a fission reactor, and it is therefore inherently safe! In a fusion reactor the nuclear fuel (e.g. deuterium) cannot create a nuclear explosion.

In a fission process the critical mass plays a crucial role. Critical mass is the minimum mass of a fissile material that sustains a violent chain reaction; this is the minimum mass in an atomic bomb. In a fission reactor, during a drastic accident, when a critical mass accumulates, a violent explosion could take place. Such a situation does not exist in a fusion reactor as the nuclear fuel used for fusion does not have a critical mass. In contrast to fission reactors, one would not have to dispose of any radioactive materials—there would be no 'dirty' waste!

5.5 Conditions for Nuclear Fusion

How can we fuse deuterium–tritium, deuterium–deuterium, or deuterium–helium-3 in order to achieve energy? These nuclei have positive charges and therefore they repel one another, preventing the light nuclei from fusing. The way to overcome this problem is to supply enough energy to these nuclei. When the nuclei collide at high speed, they overcome the force of repulsion of the electric positive charges and can fuse. This can be achieved in two ways: either by accelerating some particles into others or by heating all the particles to very, very high temperatures. The first option of accelerating some particles is very inefficient. In this case the initial energy of the accelerated projectiles is lost after having passed through the electrons of the cold target. However, if the mixture is very hot, thermal collisions will go on and on until all the available nuclei have fused.

In order to visualize the different types of fusion process described above, let's imagine a cold, snowy, winter day. We are watching a snowball fight between children being carried out in two different ways. In the first instance, the children are strictly supervised by their teachers and are told to stand still. They are forbidden to throw any snowballs. But some mischievous children manage a few fast throws. In the second instance, everyone is allowed to throw snowballs at everyone else. In the first case, the teachers standing between the children will also be hit by some of the snowballs. Therefore the number of children who will be hit is not very large. It is clear that in the second case everybody is going to be hit. In the second case we have much more effective interaction between the children. The balls are flying in all directions and the spread in their energy is equivalent to our definition of temperature. In this analogy, the mischievous children are the accelerated particles in the first case, while the teachers are the electrons of the cold target. In the second case, all the children are the plasma particles.

Chemists have known for a long time that chemical reactions between molecules are faster when they are hot. A log of wood will burn when it is heated to several hundred degrees in an ordinary furnace. Housewives know that food is ready after it has been cooked for the desired time and at the right temperature.

The number of reactions in a unit of time (say one second) is determined not only by the temperature of the plasma but also by the probability for the nuclei to fuse. As previously explained, between the positive nuclei there is a force of repulsion. The higher the temperature, the higher the energy of the ions in the plasma. The probability of overcoming the barrier of repulsion is larger at higher energies. This probability is given schematically in figure 5.7 by the graph denoted by P, which describes the penetration of the nuclear barrier. The higher the

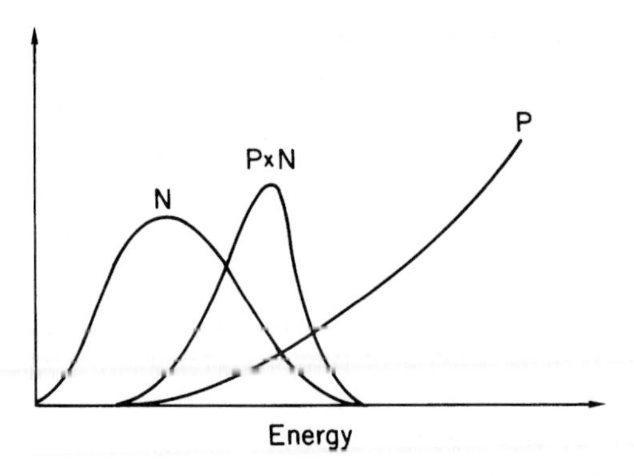

Figure 5.7 The probability for a fusion process. *P* represents the probability to overcome the Coulomb repulsion between the fusion nuclei, *N* describes the energy distribution and $P \times N$ is the total rate of the fusion process.

energy, the easier it is to penetrate the Coulomb repulsion, which is known as the Coulomb barrier. Above a certain energy the barrier is overcome. As can be seen in figure 5.7, below a certain energy there is no penetration, namely $P = 0$. In a hot plasma with a definite temperature, the particles are distributed according to the bell-shaped distribution demonstrated schematically in figure 5.7 by the letter *N*, which describes the number of nuclei for every particle energy (see also Chapter 2). For larger temperatures, the bell shape of *N* becomes wider (see figure 2.2). Figure 5.7 shows the interesting feature that the number of particles having very large energies is very small but their effectiveness in penetrating the barrier and causing fusion is very large. Therefore, although they are few, these high-energy plasma particles are of great importance for the total rate of the fusion processes. In figure 5.7 the graph $P \times N$, which is the product of the two curves *P* and *N*, represents the total rate of fusion processes. From the $P \times N$ figure one can see that the maximum rate occurs at an intermediate energy value for which the number of particles is not yet too small and their penetration of the barrier is not yet very high. This type of calculation for deuterium–tritium fusion requires temperatures of about 100 million degrees in order to get fast fusion reactions. The fusion of deuterium alone would require even higher temperatures. The actual temperature required depends on the particular fusion reaction under consideration.

The rate of fusion reactions is important in calculating how much energy we get out from the fusion process. The rate of thermonuclear reaction is dictated by the plasma conditions of density and temperature. In order to achieve the required conditions, energy has to be invested. The English scientist J. D. Lawson was the first to make a theoretical energy balance between the invested energy producing the plasma and the output fusion energy. The 'break-even' condition known as the Lawson criterion describes the equality between the invested energy and the recovered fusion energy. This criterion requires the product

$$\text{Plasma Density} \times \text{Confinement Time} = \text{Lawson Number.}$$

For deuterium–deuterium fusion the Lawson number is one hundred times larger, and the temperature is larger by a factor of five, than in the deuterium–tritium case. This is the reason why deuterium–tritium fusion appears to be the easiest to achieve among all the other nuclear fusion possibilities. The Lawson criterion expresses the importance of keeping the hot plasma together for a long time (the confinement time), which is essential to produce sufficient energy to compensate for the energy supplied initially to heat the plasma.

Let's make a detour and try to acquire a feeling for the Lawson criterion. Imagine many thousands of hens and cocks spread out over a large area. The goal of the owners is to obtain as many chicks as possible. For this purpose, it is clear that their density (how close they are together) and how long they will be confined will also determine the number of chicks produced. The temperature also plays an important role in order to keep up the family reproduction. (If it becomes too cold, this type of fusion also dies down.)

Thus far we have discussed two conditions necessary for nuclear fusion energy gain: (a) an appropriate high-temperature plasma is required in order to achieve a high rate of fusion reactions; and (b) the product of the plasma density and the plasma confinement time has to be larger than the Lawson number.

From Lawson's criterion one can see that there are two basic and distinctive approaches to fusion. One approach is to take dilute gases of hydrogen isotopes, to ionize them, heat the plasma and to confine it as long as possible, thus obtaining large confinement time and small densities so that the product is larger than the Lawson number. In this approach the plasma is confined by magnetic fields and the confinement devices are called magnetic bottles.

In the second approach one tries to compress the deuterium–tritium plasma to very high density, thousands of times larger than the solid density of material. In this case the confinement time is extremely small, since the very compressed matter explodes and diffuses. The time is fixed by the inertial properties of matter (that is, the time that the material remains in its original form) and thus this scheme is called inertial confinement fusion. The necessary densities for this approach could be achieved by using very powerful lasers or very intense particle beams. These are used to compress small pellets containing the deuterium–tritium mixture. Although this thermonuclear fusion is violent, it is small enough to be contained in a reasonable-sized vessel without damaging it. By repeating the fusion in this scheme many times one should be able to gain energy under appropriate conditions.

5.6 Ignition Temperature

A hot plasma at thermonuclear temperatures loses a considerable amount of its energy in the form of radiation. The nuclear fusion process has to produce more energy than is lost from the radiation of the plasma. This requirement determines the minimum temperature for a nuclear fusion reactor to be 'self-sustained'. Physicists have calculated the rate of energy production from fusion as explained in the previous section. The radiation (the lost energy) is created in the plasma medium by the motion of the charged particles, and, in particular, by the electrons. In a hot plasma, electrons are moving chaotically in all directions and they collide with other electrons and ions. Due to these collisions, some of the electrons lose speed and therefore energy. This lost energy of the electrons is released in the form of photons. The collection of all the photons is called radiation, or more accurately electromagnetic radiation. This mechanism of radiation production is called *bremsstrahlung* which is the result of deceleration of electrons (in the presence of ions) in the plasma. The meaning of *bremsstrahlung,* a German word, is 'braking radiation', as a result of the slowing down of electrons, i.e. applying the brakes and emitting photons. This phenomenon is also used for medical diagnosis and therapy in machines producing X-rays. A beam of electrons is accelerated by a high potential and then these electrons are focused and directed onto a target. The electrons lose speed inside the target, i.e. they are decelerated and radiation—X-rays in this case—is emitted. This radiation is similar to the radiation in a plasma. Calculations show that at lower temperatures, the radiation loss rate is larger than the rate of energy production from nuclear fusion. As the temperature increases, the production energy as well as the radiation losses increase. However, the fusion energy production increases faster than the radiation loss. There exists a temperature which is called the ignition temperature above which more energy is produced by nuclear fusion than is lost by radiation. For these temperatures a self-sustaining fusion reactor is possible.

Let's summarize the general balance of energy. It is similar to the bookkeeping of a big enterprise. First, the big enterprise has to invest a certain amount of money. In our plasma case this is the energy required to create and to heat the plasma. The sales derived are equal to or larger than the money invested to buy the goods. This condition is our 'Lawson criterion' for obtaining more energy than was originally invested. However, this is not the complete picture; we may still not have a profit. There are many losses (overhead expenses)—in the plasma case the loss due to radiation energy. In order to make a profit, it is necessary to sell more so that we gain back the investment and are able to cover our losses as well. Similar book-keeping is required with the energy balance in the fusion process. While in physics scientists have calculated and know the conditions for profit, in business this problem seems to be unsolved in general, although some know very well how to 'ignite' money and make some good profits.

5.7 Magnetic Confinement—Magnetic Bottles

Setting our imagination in full gear, let's visit a big playground with thousands of kindergarten children at play; running around clamorously, playing with different balls and colliding with one another. Picture some of them wandering off on their own without supervision. See the small feet outrunning the bigger and slower-moving ones of their supervisors. Hear some shouting and others laughing or crying. What a tumultuous situation this is indeed. Yet, unbelievably, occasionally, under very strict supervision, this big chaos can become organized in such a way that all the children are playing the same game. They can all be lined up in an orderly manner, listening to their teachers' instructions. But do you believe that children, in general, can listen or play the same game for a very long time? Of course this is very doubtful. The bedlam is bound to start all over again and the nice collective behavior of the children will 'explode' at any moment to a chaotic motion. Does controlling several thousands of kindergarten children in a big playground, with very nice and beautiful surroundings, appear to be a great task? Perhaps, but the physicist, in his laboratory, trying to contain and control millions of millions of millions of charged particles, with very strong interactions between them, inside a small vessel, is faced with a much greater task. But if he wants to achieve a hot and lasting plasma necessary to solve the energy problem, then this is what he has to cope with daily in his laboratory.

Let us assume for the time being that the physicist in his laboratory knows how to produce and heat the deuterium–tritium plasma to 100 million degrees or more. In order to gain energy the plasma has to be confined (recall the Lawson criterion) and the temperatures should be as hot as the center of the Sun. But how can a small man-made sun be bottled? Plasma, unlike any ordinary gas, is a good conductor of electricity and its motion is influenced by magnets. Very hot plasmas cannot be insulated by regular materials such as rock wool (mineral wool made by blowing a jet of steam through molten rock such as limestone or siliceous rock or through slag and used chiefly for heat and sound insulation). This insulation has to be provided by powerful electromagnets. There is *no* material that can come in contact with a plasma at a temperature of millions of degrees Celsius without itself becoming plasma. The only way to detach a plasma from the walls of the vessel is by the use of magnetic fields. The plasma particles are electrically charged and therefore they are usually trapped around magnetic field lines (see Chapter 2).

Charged particles, like the electrons and the nuclei in the plasma, have difficulty in crossing the lines of force of the magnetic field. Therefore a plasma can be confined by different magnetic field configurations. These magnetic fields keep the plasma inside a 'bottle' so that the electrons and the ions do not touch the walls. This is the principle of the magnetic confinement scheme and the device that confines the plasma is called a 'magnetic bottle'. (Of course these are not the ordinary bottles that we are familiar with.) The term magnetic bottle was coined in the 1950s by Professor Edward Teller, one of the pioneers of thermonuclear fusion physics.

The motion of the particles in a plasma exerts a pressure in a similar way to the pressure of a gas. The thermal pressure in the plasma is due to the particle motion. For higher temperatures, the motion is more vigorous, causing a higher pressure. The number of plasma particles is also a factor in determining the pressure; for larger densities one has higher pressures. Usually, the plasma pressure, like the pressure of a gas, causes the plasma to expand. The plasma can be confined and not allowed to expand into the wall by a magnetic field. This phenomenon can be understood by the fact that the magnetic field also has a pressure. The balance of pressure can be written as

Magnetic Pressure Outside Plasma

= Thermal Pressure + Magnetic Pressure Inside Plasma,

where the magnetic pressure is proportional to the square of the magnetic field. Therefore the magnetic field outside the plasma exerts a pressure on the plasma opposite to the particles' thermal pressure.

Scientists define the ratio of the thermal pressure to that of the magnetic pressure inside the plasma by the Greek letter beta. (This beta has nothing to do with the radiation discussed in Chapter 1.) For magnetic confinement devices, beta is usually smaller than unity—that is to say, the magnetic pressure inside the plasma is larger than the thermal pressure.

5.8 Plasma Diffusion

The presence of magnetic fields in a plasma causes the electrically charged particles to follow spiral paths encircling the lines of magnetic force. The negative electrons spiral in one direction while the positive ions spiral in the opposite direction. Consequently, the charged particles are not free to move across the magnetic field lines (from one magnetic line to another). One can imagine that each particle is tied around a magnetic line of force and if these magnetic lines do not hit the wall the particles will be confined inside the vessel. The radius of the spiral motions, the so-called gyration radius, depends also on the particle velocity and mass. Therefore, the electrons will move in a much smaller spiral than the ions. Since the particles in the plasma do not move with the same velocity (remember the bell-shaped distribution; see Chapter 2) neighboring electrons might have different speeds, thus causing them to spiral with different curvatures. As a result, a plasma particle will occasionally collide with other particles. These collisions cause the electrons and ions to move from one magnetic line of force to another. In this way the charged particles can move across the magnetic lines of force and eventually escape and hit the wall. The motion of the electrons across the magnetic lines of force is called plasma diffusion.

Let's take our runners to a field track. As long as each runner remains in his own lane and there are no collisions, the children will run around the track without falling inside the field or outside the fence. However, if there are many, many, children, even though they try very hard to stay in their own lanes, there are bound to be some collisions. Due to these collisions, some of the children will move from lane to lane and eventually reach the fence just as the plasma expands to the wall.

Perhaps a more pictorial scene can be envisaged when watching car racing. As long as all the cars are racing in an orderly manner they will all remain inside their tracks (confined inside the vessel). But as we know, in car racing, there are also undesired collisions (as in the plasma). As soon as one car collides with another car and another car with still another one, the colliding cars are forced to leave their tracks and to migrate to the sides just as the plasma expands to the wall.

There are other reasons why a charged particle can move across the magnetic field lines. The radii of curvature of the trajectories of the charged particles of the plasma depend on the strength of the magnetic field. The larger the magnetic field, the smaller is the radius of the spiral of the electron and ion motion. In a magnetic confinement configuration, the magnetic field is not the same everywhere. There are regions with more lines of force than others. The change of magnetic field over the course of a spiral causes the particle to change its radius. If the magnetic field is uniform, the spiral path of the electron will have a constant radius. For larger fields, the radius is smaller. Therefore, for the situation shown in figure 5.8 the radius of curvature of the electron's path is smaller below the line, where the magnetic force is stronger, and larger above the line in

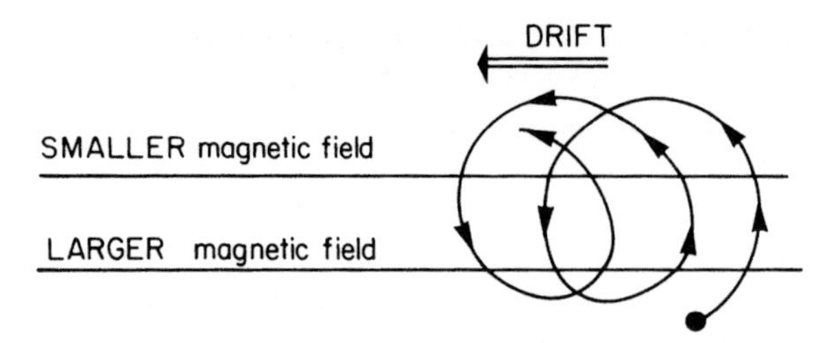

Figure 5.8 A trajectory of an electron drift due to an inhomogeneous magnetic field.

the area of weaker magnetic strength. In this way the electrons can drift and reach the walls of the containing vessel. This kind of diffusion causes the electrons to drift in one direction while the ions cross the magnetic lines in the opposite direction. The reason is that charged particles with opposite signs (negative and positive) spiral in opposite directions. The diffusion of the negative electrons in one direction and the positive nuclei in the opposite direction causes a separation between positive and negative charges. As described in Chapter 2, these charge separations introduce an electric field. This electric force combined with the magnetic force causes the plasma as a whole to move from strong magnetic fields to lower magnetic fields. This type of diffusion can be summarized as follows: Due to the inhomogeneous magnetic field, electrons diffuse in opposite directions to the ions. Consequently, an electric field is created in the plasma. The cross contribution of this electric field and the magnetic lines of force cause the plasma to drift. This drift can push the plasma towards the walls. In order to overcome the diffusion of plasma towards the wall, specially shaped magnetic fields are necessary. Calculation of the details of the diffusion of plasma is one of the most difficult problems and issues to be overcome in order to achieve a controllable nuclear fusion reactor in the future.

5.9 Plasma Instabilities

In Chapter 2 we learned that the plasma particles behave in a collective way. Different types of wave are created in the plasma due to the collective behavior of the electrons and the ions. The types of wave that can be created in a plasma are numerous and we shall not discuss these possibilities. Every collective motion can be stable or unstable. For example, suppose a small displacement takes place in a plasma confined by a magnetic field. If the collective motion behaves in a stable way, the displacement will be restored to the original motion. However, if the collective motion is unstable, any small displacement increases rapidly and the motions of this collective phenomenon break up.

These plasma instabilities are characterized in two categories called macroinstabilities and microinstabilities. Macro means large and micro means small. Whereas the macroinstabilities are on a large scale inside the plasma, the microinstabilities are on a small scale relative to the size of the plasma. Macroinstabilities can lead to a complete loss of confinement. Microinstabilities usually increase the drift of the electrons and the diffusion of the plasma. These instabilities can occasionally cause a chaotic motion in the plasma, resulting in a loss of confinement.

One can compare the prediction of plasma instabilities with the weather forecast. As we all know, the prediction of the weather is not always accurate in spite of the fact that there are many satellites and devices measuring the motion of our atmosphere. One of the main reasons that the weather is sometimes unpredictable is the fact that a small local perturbation can amplify very quickly and change completely the picture previously observed by the satellites. Scientists use statistics and probabilities (see Chapter 2) to describe these situations. However, an absolute and detailed prediction so far seems impossible. From this point of view, a plasma behaves like the weather. A small perturbation in the plasma density, for example, can increase and an instability will develop in such a way that plasma confinement might be lost; the plasma runs to the wall! However, there is a basic difference between the weather and plasma instabilities. Whereas the weather cannot be controlled the plasma in the laboratory will become controllable.

Understanding plasma instabilities is vital in confining and controlling a thermonuclear plasma. Over the last 30 years, as a result of extensive research programs and international collaboration between experimental and theoretical plasma physicists, considerable progress has been made. There is now a general feeling that almost all of the plasma instabilities are understood theoretically and these instabilities can be controlled in a fusion reactor. However, stability is not an independent issue. It is important to have a stable plasma with the relevant parameters for fusion devices. For example, the value of beta (the ratio between the thermal and magnetic pressures; see Section 5.7) should be large enough to allow one to build an economic 'magnetic bottle'. The rate of energy production is proportional to beta. Thus beta is used to measure the *efficiency* of energy production in a magnetic configuration. Although a beta of a few per cent might be enough to ignite a plasma, larger values are needed for economic purposes. The beta value is strongly connected with plasma stability. Plasma theoreticians have calculated two regions of stability: low beta (a few per cent) and high beta (40 to 50%). The low-beta stability regime has already been achieved and scientists are now aiming to achieve high-beta stability.

5.10 Plasma Formation

The simplest way to create a plasma is to pass a high-voltage electrical discharge through a gas. This is achieved in fluorescent neon lamps. The gas inside the neon lamp is in a plasma state of matter. This plasma is a cold plasma created by the electric discharge in the following way. An electron current flows between two electrodes from one end to the other end of a vacuum tube. The gas inside this tube is diluted and the streaming electrons collide with the molecules of the gas. During these collisions, the bound electrons (of the molecules and the atoms) are excited to higher energy levels and emit photons when returning to their lower energy levels. These photons are the radiation that we see in neon lamps. In the more advanced fluorescent lamps, photons emitted from the plasma gas inside the tube hit the glass envelope, which absorbs those photons and radiates more photons of lesser energy. This radiation is the visible light that we see.

It is also possible to create a plasma outside the magnetic bottle and to inject it into the reaction chamber. In this case a plasma is produced by an electrical discharge, as described above. The electrons and ions are accelerated into the magnetic bottle by a combination of electric and magnetic fields. In a working thermonuclear fusion reactor, the temperature is very high. Therefore, the deuterium–tritium or deuterium can be injected inside the magnetic bottle in the form of a gas. Since the temperature in the reactor is so high, the gas will at once be ionized. In this case the plasma is formed by pure heating.

5.11 Plasma Heating

In order to start a nuclear fusion reactor, one has to heat the plasma to many millions of degrees. There are a few ways to invest the necessary energy into the plasma state. The first way is to pass a large electric current through the plasma. Since the plasma is an electrical conductor, it is possible to pass a current through it. This heating is called 'ohmic heating' because it depends on the resistance of the medium that is carrying the current, measured usually in units of ohms. This type of heating is exactly the same as heating water with an electric heater. The only difference is, of course, the size of the current under consideration. For magnetic confinement devices, the currents are much larger than the necessary current for boiling water. One of the problems with the 'ohmic heating' in a plasma is the fact that, as the temperature increases, the plasma resistance becomes smaller and therefore the heating is less effective. As we learned in Chapter 2, when the temperature is increased the collision frequency in the plasma decreases. The plasma resistance is proportional to the collision frequency and thus an increase in the temperature will result in a decrease of the resistance. Therefore it is not possible to heat the plasma to the desired thermonuclear temperatures. Consequently, other schemes for heating are necessary.

The second possible mechanism of heating is by 'magnetic compression'. In general, a gas can also be heated by a sudden compression. Let's imagine that many people are suddenly transferred from a very spacious room to a very small one. Obviously, it will become much hotter in the second room due to the compression. The magnetic field exerts a pressure which confines the plasma. By increasing the strength of the magnetic field, the plasma is compressed and consequently heated. This process has an additional advantage since in compressing, i.e. squeezing, the plasma, the nuclei are forced closer together and thus the fusion process is increased. It is still uncertain how effective and relevant this scheme will be in igniting a plasma.

Another procedure to increase the temperature of the plasma is by injecting energetic neutral particles into the vessel. Neutral particles (i.e. atoms) cannot be accelerated to high velocities in a direct way. An electric force acts only on a charged particle, and so only charged particles can be accelerated. The injection of neutral beams into the magnetic bottle is, for that reason, done in an indirect way. First, deuterium ions, which have positive charge, are created and accelerated to the desired energies. Then, before entering the magnetic bottle, these ions pass through a chamber containing neutral deuterium gas. The energetic ions of deuterium strip electrons from the neutral atoms of the gas and in this way become neutral. This process is called 'charge exchange'. Consequently, a stream of high-energy neutral deuterium atoms is formed and enters the magnetic bottle to heat the plasma. Inside the magnetically confined vessel, the neutral stream of deuterium atoms collides with the plasma, and the atoms become ionized and trapped within the magnetic field. The high-energy beams transfer part of their energy to the plasma particles during collisions and in this way the plasma temperature is increased. The above scheme of heating is rather complicated and confusing to a non-scientist. In order to simplify the explanation, one has to remember the main issue, namely, that an energetic beam of particles is used to heat the plasma. The particles are first ions, which eventually are transformed into neutral atoms before entering the plasma vessel. Once inside the vessel, they collide with the plasma and are transformed again into ions. During the collision they transfer their extra energy to the plasma and in this way provide heat. This neutral beam heating is useful as an addition to ohmic heating in order to ignite the plasma.

The above complicated scheme can be simplified by the following example, which is not to be taken too seriously. A man (the charged ion) chases (is accelerated) after a married woman (the electron belonging to another ion) until he catches her (a neutral atom). They run away together (energetic neutral atom) until they tire of one another and thus become separated (plasma).

The fourth way to heat a plasma is by irradiating electromagnetic waves inside the plasma. This process is similar to a microwave oven where the food is heated by electromagnetic waves. The main difference is in the quantity of radiation one needs to heat the plasma. The energy from the waves is absorbed by the charged particles in the plasma which collide with other plasma particles, thus increasing the temperature of the bulk plasma. This heating can also be used as an addition to the ohmic heating when igniting the plasma.

All or some of the above heating schemes are necessary to start a nuclear fusion reactor. In an operating thermonuclear fusion reactor, part of the generated energy will serve to maintain the plasma high temperature as fresh deuterium and tritium gases are introduced. Therefore, the above schemes of heating will not be necessary for a working reactor. In present magnetic bottles, the fusion energy is not sufficient for igniting the plasma to nuclear reactions. Consequently, the present magnetic bottles operate in short pulses and the plasma must be heated again in every pulse.

5.12 The Tokamak

Many different magnetic confinement schemes have been suggested and studied during the last three decades. The scheme which is still receiving great attention in the magnetic fusion energy program is the Tokamak (see figure 5.9). This device was invented in the Soviet Union by the physicist Andrei Sakharov (a co-inventor of the Soviet hydrogen bomb and later a recipient of the Nobel Peace Prize) and was developed and built in the laboratory under the leadership of the Soviet physicist Lev Artsimovich. The name Tokamak is an acronym for 'toroidal magnetic chamber', or more accurately the acronym for 'torus', 'chamber' and 'magnetic' in Russian. The Tokamak is a device which has a hollow doughnut-shaped vessel (a torus). Imagine bending a solenoid into a circle and joining the ends so that a doughnut is created. The plasma is confined in this vessel by the magnetic lines of force which twist inside the torus. There are two major components of the magnetic field (see figure 5.10): (a) the *toroidal* magnetic field which is in the direction of the main axis of the torus and is generated by electric currents flowing in turns or rings around the torus; and (b) the *poloidal* magnetic field component which turns around the torus and is generated by the current flowing through the plasma. The combined magnetic lines are in a spiral form (helical field) turning around inside the doughnut. Many

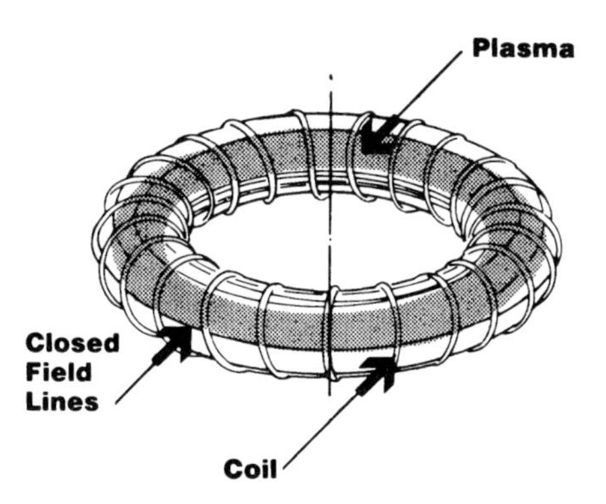

Figure 5.9 Diagram showing a Tokamak fusion design, which uses magnetic fields to trap hot plasma inside a doughnut-shaped container. (Courtesy of Lawrence Livermore National Laboratory, California, USA.)

large transformer cores are linked through the hole in the torus and, as the current rises in these primary coils, the plasma is driven around the main axis of the torus. This is similar to a transformer where current from one circuit is transferred into a second circuit. In a large Tokamak the electrical currents in the surrounding copper coils (external circuit) produce a magnetic field inside the plasma which is about one hundred thousand times stronger than the magnetic field of the Earth (which deflects a compass needle to the north). The current in the external circuit induces a current inside the plasma of the order of one million amperes (in the largest Tokamaks). Such currents are a few hundred times larger than the currents carried by the large transmission lines across the USA. As mentioned above, the external circuit induces a current in the plasma

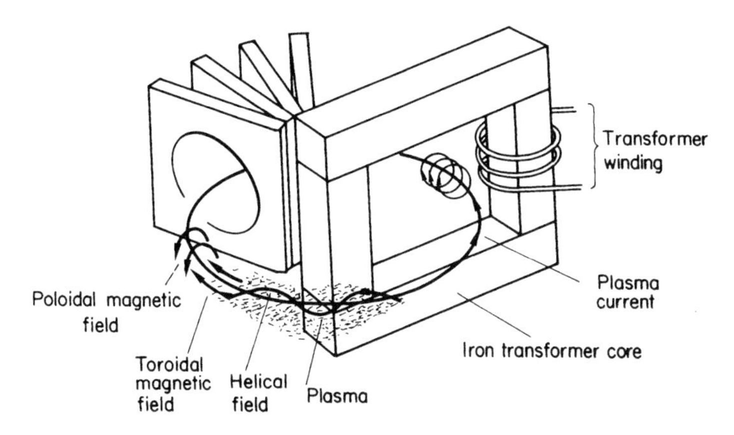

Figure 5.10 A magnetic field line in a Tokamak device.

which circles inside the torus. This plasma current creates the poloidal field in exactly the same way as any current creates a magnetic field in a circle around the current. This magnetic field confines the electrons and the ions inside the vessel and does not allow the plasma to run to the wall.

However, not all the plasma is kept completely confined and some does escape to the wall. Even if only a very tiny fraction collides with the wall, the plasma cannot be heated to the desired high temperatures. This is because, when the hot plasma hits the wall, atoms from the wall are knocked out and enter the plasma. These atoms are 'impurities'—not the hydrogen isotopes or their fusion products, but rather heavy atoms with many electrons surrounding their nucleus (the material of the vessel). The bound electrons absorb part of the invested energy and thus cool the plasma. This problem was solved by building 'limiters'—heavy bars of graphite (composed of carbon atoms) protruding from the vessel wall at some particular place. Since in a Tokamak the electrons and ions run around the torus as in a racetrack, those plasma particles which approach the wall hit the limiter. The plasma collisions with the wall are avoided all the way through the doughnut-shaped vessel. Although some of the limiter atoms will enter the plasma, they cool the plasma less effectively than the heavy atoms of the wall material (or of the first limiters built from tungsten or molybdenum). This is because the number of bound electrons in carbon (the limiter material) is small (carbon has six electrons). Moreover, after all the carbon electrons are ionized the carbon does not cool the plasma any more. Light atoms lose their electrons much more easily than the heavier elements.

One of the requirements for a reactor is that it should work continuously. Since Tokamaks use transformers in order to produce the current causing confinement and heating, they can run only for a short period. A transformer can store current driving power which is measured in volt-seconds. After the volt-seconds of the transformer are exhausted it ceases to function until it is recharged. During the recharging period the Tokamak does not operate, so that the use of a transformer is very inconvenient. To switch on and heat a Tokamak every few seconds is not only difficult but also not economic. However, research at Princeton and Massachusetts Institute of Technology in the USA has proven that it is possible to use powerful microwaves in order to drive the currents in the Tokamak. Therefore, during the recharge process of the transformers the microwaves can do the job in such a way that the Tokamak is run continuously.

The scientific landmark, where the nuclear fusion energy equals the energy invested in the heating of the plasma, the so-called break-even demonstration, was accomplished in the 1990s in the large existing Tokamak devices in USA and Europe.

5.13 Magnetic Mirrors

If an electric current flows through a solenoid, a magnetic field will be produced along the axis of the solenoid. The strength of the magnetic field is larger for denser solenoids, as more coils per unit length create more magnetic field lines. The simplest configuration of a magnetic mirror consists of a coil wound around a straight tube where the number of coils is larger at the ends than in the middle of the tube. If the same current is passed through each coil, then the magnetic field is stronger at the ends than in the middle. The lines of the magnetic field are closest (pinched) where the field is strongest. This configuration of magnetic lines slows the charged particles approaching the ends of the tube and under certain conditions the charged particles of the plasma turn back from the ends of the tube. Therefore, the ends of the tube where the magnetic field is larger act as a mirror and cause the electrons and the ions to be reflected back inside the tube.

As an electron spirals about a magnetic line of force and approaches the end region where the field is stronger the spiral path becomes smaller and smaller. For larger magnetic fields, the radius of the spiral is smaller. If the magnetic field in the mirror region is strong enough, the electron's motion along a magnetic line will stop and then reverse back into the tube. This electron will spiral back and forth between the magnetic mirrors which are at both ends of the tube. However, not all the electrons are reflected from the mirrors. Those particles having a velocity exactly along the magnetic line of force do not spiral at all and they leave the tube. In general, there is a relation between the direction of motion of the electrons and the strength of the magnetic field inside the mirror and always some of the plasma particles will leak through the mirrors. The central region of this configuration acts as a 'magnetic well', so that one can imagine that the plasma particles are in a well. While some of the particles are caught inside the well, many other particles can escape outside the well.

More sophisticated and ingenious schemes of magnetic configuration have been constructed in order to confine the plasma in the tube vessel. Plasma confinement systems have been constructed from currents passing through coils which are wound in the form of a baseball seam. The central region of the baseball-seam coil is the well and the magnetic strength increases outwards in all directions.

With even the most sophisticated magnetic mirror configurations, magnetic mirror bottles still exhibit plasma losses that escape the mirrors. The radial diffusion of the plasma towards the wall of the tube is also a very important issue and a hard problem to solve. Mirror configurations are very interesting and simpler than other configurations; however, it seems that plasma cannot be confined in a magnetic mirror device for very long periods of time. For the time being it is believed that the magnetic mirror is not the best magnetic bottle, thus this line of research is no longer being pursued vigorously. See Section 5.19 for the main magnetic fusion achievements in the past 10 years and also for other possible magnetic configuration devices.

5.14 Nuclear Fusion Reactors

During the nuclear fusion of deuterium and tritium, a nucleus of helium is created together with one neutron. About 80% of the released energy is carried away by the neutrons while the remaining 20% is contained in the helium nuclei. The neutrons do not have any electrical charge. They are neutral and therefore magnetic and electric forces do not affect them, allowing them to escape from the plasma. As the helium nuclei, produced during the fusion interactions, are charged particles and are affected by the magnetic forces, they remain in the plasma chamber. Their energy is used to ionize and to heat the incoming deuterium–tritium gas, enabling the fusion to continue on a steady-state basis.

The doughnut plasma chamber is surrounded by a blanket of lithium in order to produce tritium. The tritium is removed as a gas from the lithium blanket and is returned into the reaction chamber together with the deuterium gas. While producing tritium, the energetic neutrons deposit their energy in the lithium blanket causing the lithium to be heated to high temperatures (more than 1000 °C). Inside the heated lithium blanket pipes are placed to allow the flow of some liquid or gas (such as potassium or helium). This liquid boils and the hot vapor is used to drive a turbine generator for the production of electricity.

To summarize, the fusion of D–T produces helium and neutrons. The helium nuclei heat the plasma and keep the thermonuclear fusion alive. The neutrons produce tritium, as well as supplying the energy for heating a liquid or a gas. This liquid or gas runs a turbine in order to produce electricity, just as in any other electrical power plant.

The strong magnetic fields which confine the plasma require large amounts of electric power in order to maintain their operation. This power supply can be reduced by using superconductor magnets. A superconductor material has the unusual property of carrying currents without any resistance, thus saving electrical power. Until recently, these superconductor metals and alloys operated at very, very low temperatures, usually about 263 °C below the freezing point of water. (There is an absolute minimum temperature of 273 °C below the freezing point of water. Nothing in the Universe can exist below this absolute zero point of temperature.) These superconductors work in a very, very cool environment. Scientists have produced superconductors which operate at much higher temperatures. These superconductors operate above the freezing point of nitrogen (liquid nitrogen) and therefore these temperatures are technologically easier to achieve. This in itself is a revolution in electricity and in the research of matter. From the nuclear fusion view, one of the major contributions of this new discovery will be the building of supermagnets necessary to produce large magnetic fields for plasma confinement.

In figure 5.11 the International Thermonuclear Experimental Reactor (ITER) computer-generated model is shown. ITER is a Tokamak configuration designed to burn deuterium–tritium for 20 minute periods and to reach a fusion power of 1.5 gigawatts. This power is comparable with a standard generating plant. All the general characteristics necessary for a workable nuclear fusion reactor, as described in this section, are incorporated in the ITER design. The ITER Tokamak is based on the 'state of the art' plasma physics knowledge of the second millennium.

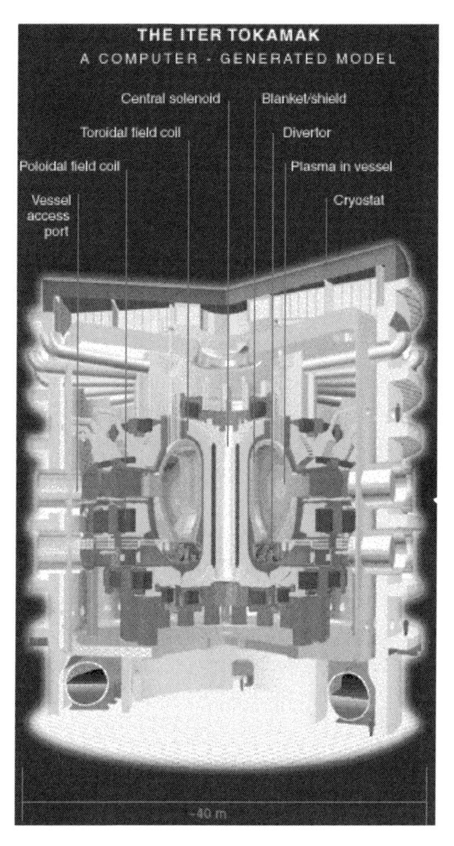

Figure 5.11 The ITER Tokamak, designed to reach a fusion power of 1.5 gigawatts. (Courtesy of the International Thermonuclear Experimental Reactor Project.)

5.15 Inertial Confinement with Lasers

The principle of the laser, namely the idea of stimulated emission, had already been suggested in 1917 by Albert Einstein. However, it was only in 1960 that the American scientist T. H. Maiman constructed the first laser system. In 1964 the American scientist Charles H. Townes and the Soviet physicists Nikolay G. Basov and Aleksander M. Prokhorov were awarded the Nobel prize in physics for work done in the 1950s leading to the construction of the laser.

Lasers are devices that produce intense beams of light of a single color, i.e. a single wavelength. The name laser is an acronym for light amplification by stimulated emission of radiation. However, people have gotten into the habit of referring to the device that creates this light as the laser. Different lasers are named after the materials used as the medium to amplify the light. For example, a solid-state laser or a gas laser is one in which the amplifying medium is a solid or a gas. More specifically, a neodymium laser is one that uses the material neodymium in the medium where the light is amplified by stimulated emission.

Intense laser light beams can vaporize and produce plasma from every material here on our planet Earth including the hardest ones. Lasers are used today in medicine for eye surgery and for other microsurgeries. Lasers are used in industry for drilling and cutting accurately. More and more lasers are being used in medicine and industry today.

The laser principle is based on the idea that excited atoms or molecules can radiate in an orderly way. A necessary condition is to have a 'population inversion', where there are more electrons in an 'upper' state than in a 'lower' state. Usually, all the electrons in equilibrium tend to stay in their lowest energy levels. By heating these atoms, or by investing energy into the medium in which they reside, some of the electrons will jump to excited states. While these electrons are returning to their lower energy states, they emit radiation. This spontaneous radiation is the usual light we see from incandescent bulbs or other irradiating objects. However, if the excited electrons jump back to their ground state in response to some 'external order' then the emitted photons are in phase (in step) with one another. Such an 'order' is given by other photons which have exactly the same energy as the energy available in the excited states. The new emitted photons cause other excited electrons to 'jump down' together in such a way that the radiation is amplified (see figure 5.12). The laser light is coherent, meaning that it has a single frequency (i.e. color) and all the photons are in step with each other.

A simple explanation showing the difference between laser light and the usual light from a bulb can be shown by the following illustration. Imagine a huge number of soldiers standing on tables of equal height. The commander calls out two orders: (1) Everyone can jump down

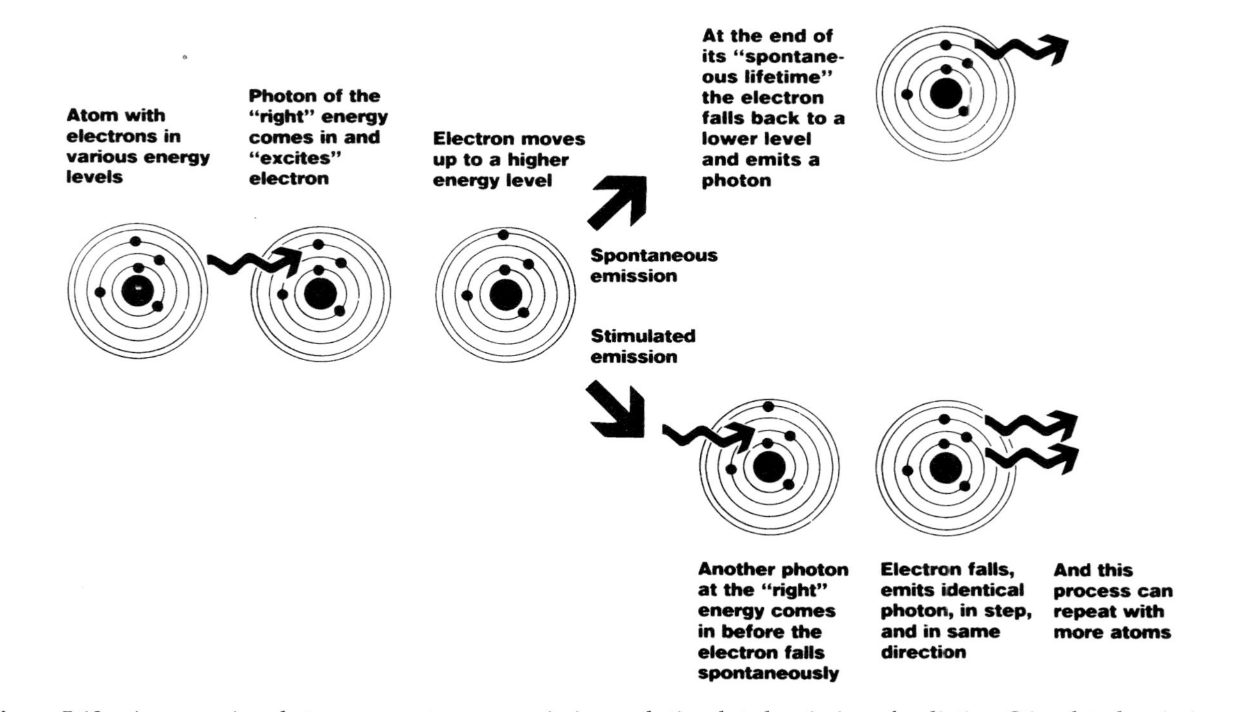

Figure 5.12 A comparison between spontaneous emission and stimulated emission of radiation. Stimulated emission of radiation is the process which gives a laser its special characteristics. (Courtesy of Lawrence Livermore National Laboratory, California, USA.)

whenever he feels like it during the next 10 minutes. (2) Everyone has to jump down after the commander counts to three. Case (1) is analogous to the regular light bulb whereas case (2) is the 'laser effect'.

Remember the Lawson criterion: 'plasma density × confinement time = Lawson number'. There exists the possibility of a very small confinement time and very large densities. These densities are about one thousand times larger than the density of water. How can one squeeze the plasma to such high densities? The gas of deuterium and tritium is inserted into a small pellet of glass, or other appropriate material, less than or about one millimeter in size. This pellet is heated by a laser very rapidly to thermonuclear temperatures. Lasers have the advantage of being easily focused onto very small spots (see figure 5.13). Moreover, the intensity of the laser radiation can be very large. Some of the laser energy is absorbed by the pellet and a plasma is formed from the outer surface material. This plasma flows off, like the gases of a rocket, and causes the remains of the pellet to move very rapidly inwards. If the deuterium–tritium mixture at the center of the pellet is highly compressed, then the thermonuclear fusion reactions occur very rapidly. The tremendous pressures generated by the fusion energy explode the pellet. A new deuterium–tritium pellet is then introduced inside the vacuum chamber where the process is repeated again and again. Most of the nuclear fusion energy is carried by the neutrons which are absorbed by some liquid. This liquid is heated by the neutron energy and then is removed in a convenient manner to operate a turbine generator which produces electricity.

Very powerful laser beams are required for the purpose of inertial confinement fusion schemes. In order to build such powerful lasers one begins with a small laser of low power, usually called an 'oscillator'. The laser pulse runs inside the oscillator between two mirrors, until it builds up to the desired lasing conditions. At an appropriate time one of the mirrors becomes transparent and the laser light leaves the oscillator. This energy is amplified when passing through a laser medium without mirrors. The latter is called an amplifier. Using beam splitters the laser pulse is split into several beams of laser light. Each beam will pass through a set of amplifiers, increasing its energy. The amplified laser beams emerge from their amplifiers at the same time and through an array of lenses are focused onto the pellet in a symmetric way. The many laser beams hit the pellet from different directions at the same time, compressing it inwards.

The physics of the interaction between high irradiance lasers and matter, leading to high compression causing a thermonuclear burn, can be summarized schematically by the following intercorrelated sequence of events (see figures 5.14(a), (b), (c) and (d)):

$$\text{Laser Absorption} \rightarrow \text{Energy Transport and Compression} \rightarrow \text{Nuclear Fusion} \rightarrow \text{Energy Release.}$$

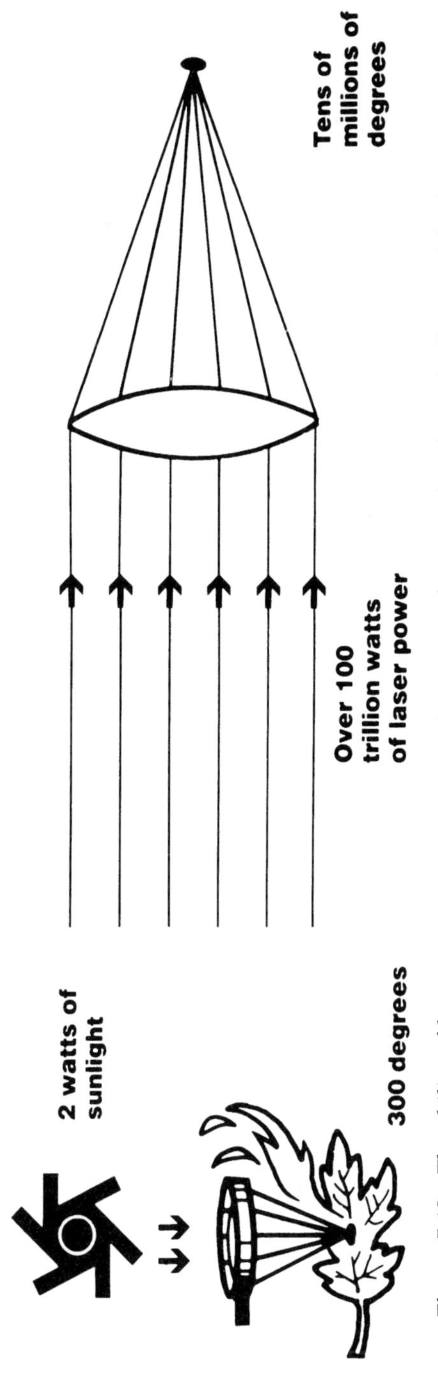

Figure 5.13 The ability of lasers to concentrate energy in space and time makes them ideally suited for fusion experiments. (Courtesy of Lawrence Livermore National Laboratory, California, USA.)

(a)

(b)

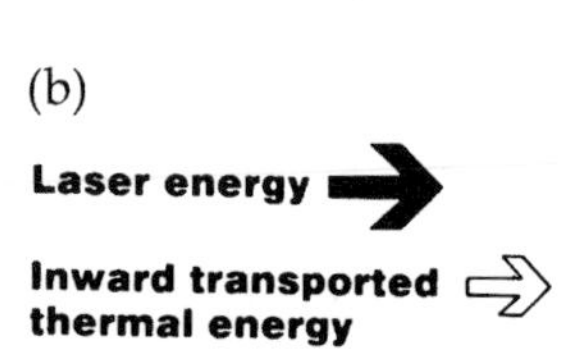

Figure 5.14 (a) Laser absorption by a pellet. (b) Energy transport and the compression of a pellet. (c) The nuclear fusion process. (d) The central fuel ignition causes the fusion burn to propagate throughout the fuel and releases energy. (Courtesy of Lawrence Livermore National Laboratory, California, USA.)

Step 3: when the fuel reaches a temperature of 50 million degrees and a density equal to 20 times that of lead, the fuel begins to fuse (burn) at the center of the target.

(d)

Step 4: the central fuel ignition causes the fusion burn to propagate throughout the fuel, producing as much as several hundred times more energy than is delivered by the laser beams that started the fusion reaction.

The laser radiation is absorbed in the outer periphery of the plasma, up to some critical surface where the laser cannot penetrate further. The critical surface of the plasma is like a mirror to the laser light and it is determined by the plasma density. When the plasma is created by a laser beam the density varies. It increases gradually from smaller densities in the outer surface (corona) to higher densities inside the target (pellet). The laser consists of photons of light which are absorbed by the electrons in the plasma. This process is called 'inverse bremsstrahlung', because photons are absorbed and electrons are accelerated.

Laser light is an electromagnetic wave which has a wavelength and a frequency of oscillation. The laser electric and magnetic fields (like any other electromagnetic fields) oscillate like the motion of the waves in the ocean. The direction of these oscillations is perpendicular to the direction of the laser irradiation. Inside the plasma, there are also waves which oscillate. These are the collective phenomena of the electrons and the ions as explained in Chapter 2. At the critical surface the frequency of the plasma wave equals the frequency of the laser light. In other words, the number of collective oscillations that the electrons perform in a period of time (a second) is equal to the number of oscillations of the electric field of the laser during the same period of time. The plasma critical surface plays a special role in the physics of laser–plasma interactions.

Let's imagine a forest. When we first enter the forest the trees are scattered. As we go deeper into the forest, the number of trees increases and the trees are spaced closer together. In such an imaginary forest, we eventually reach a point where the trees are so close together that it is impossible to pass through and we are forced to turn back. This dead end in our forest is our critical surface. The critical surface depends on the density of the trees just as in the plasma case where the critical surface is fixed by the electron density. Let's send a group of people and independently a group of elephants into this forest. It is obvious that the people will be able to penetrate much deeper into the forest than the elephants. When we use two beams of different wavelengths, the critical density changes accordingly, namely, the laser with the shorter wavelength will penetrate a higher plasma density than that of longer wavelength.

The energy absorbed by the electrons inside the plasma up to the critical surface is transported inwards to the ablation surface and outwards into the expanding plasma. The ablation surface, another important surface in this game, is the place where the target material is evaporating and the plasma is created. It is like the place in a rocket where the exhaust gases emerge and force the rocket to move in the opposite direction. In our case, due to the exhaust of the plasma material from the ablation surface outward, the inner core of the pellet is pushed inward. In this way the center of the pellet, the core, is compressed. One of the difficulties in laser–matter interaction is the fact that the process of energy transport from the critical surface to the ablation surface, where plasma is created, is not efficient. There is a transport inhibition of the absorbed energy. For high irradiance lasers only a few per cent of the absorbed energy is transported inside the pellet while most of the laser energy is wasted.

Because of the above difficulty, physicists are trying more sophisticated schemes, in particular to convert the wavelengths of the large laser systems, which are in the infrared regime, to very short-wavelength irradiation. These schemes are called indirect drive, because the lasers hit a target which produces X-rays. This short-wavelength light might compress the core of the pellet in a more efficient way. Returning to our imaginary forest, one can visualize that the shorter wavelength (the people) will penetrate further than the longer wavelength (the elephants). Thus the energy is deposited nearer to the ablation surface (where the plasma is created) for the shorter-wavelength case. The ideal case is to deposit the laser energy at the place where the plasma is created. This can be achieved with X-ray radiation.

The crucial step of the inertial confinement approach is the compression stage. The center of the pellet has to be compressed to very high densities before it is heated. It is easier to compress a cold material than it is a hot one. (It is easier to coax a crowd into an air-conditioned small room than into an overheated small room. In both cases there is no desire to be squeezed into a small room; however, given the above two choices the air-conditioned room would be preferable.) It is very important that the pellet is heated from the outside inwards (like cooking a hamburger or a steak on a barbeque) since it is easier to compress a cold material than a hot one.

Due to the sudden strong heating of matter in the absorption region, high pressure is exerted on the surrounding material which leads to the formation of intense shock waves moving into the interior of the target. This high pressure, together with the blowing off of the ablated material, drives the shock wave. A shock wave is a strong pressure wave, such as produced by a supersonic aircraft, or during a lightning discharge. (Thunder is the sonic boom produced by lightning.) Shock waves have a sharp wave front where compression takes place. Shock waves travel faster than ordinary acoustic waves. They compress the matter that they are passing through. In inertial fusion pellets, convergent compression waves (a sequence of shock waves which join together at the center simultaneously) significantly reduce the laser energy needed for laser fusion.

The structure of the target is crucial in achieving high compressions. The subject of pellet structure for high compression is still an open question. Sophisticated pellets usually contain multilayers or multishells.

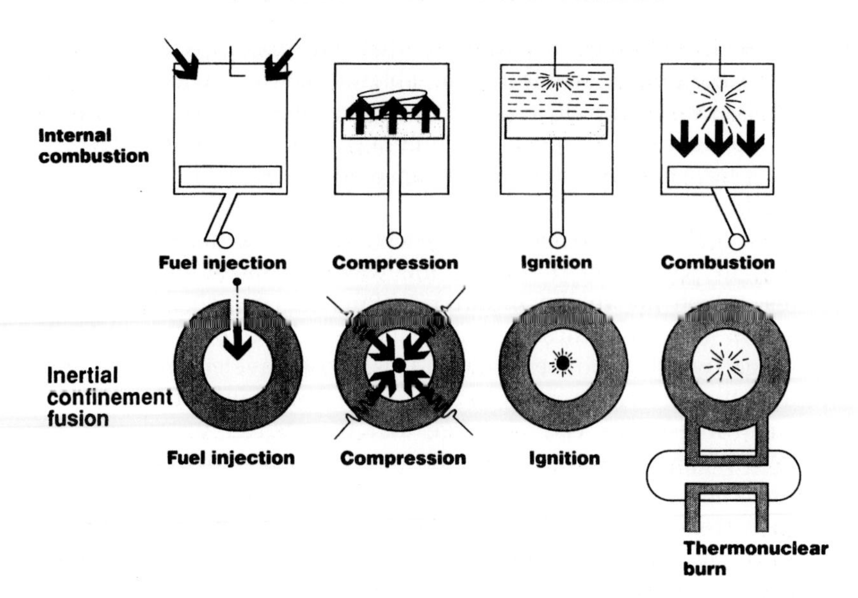

Figure 5.15 An analogy between an internal combustion engine and an inertial confinement fusion reactor. (Courtesy of Lawrence Livermore National Laboratory, California, USA.)

Once the core of the pellet has reached the high densities required, thermonuclear reactions start at the center of the pellet. The pellet burns from inside producing nuclear fusion energy. The pellet is compressed and heated from the outside inwards and it burns (nuclear) from the inside outwards.

To summarize the inertial confinement power plant, one can make an analogy with an internal combustion engine (see figure 5.15). In both cases there are four stages: fuel injection, compression, ignition and finally combustion for the engine case and the thermonuclear burn for the inertial confinement case.

The purpose of an inertial fusion power plant is to convert nuclear fusion energy into electric energy. The thermonuclear output of helium nuclei and neutrons together with the debris from the fuel pellet (and the emitted X-rays from the processes involved) are converted into heat which in turn is converted into electricity. The main structure of the reactor includes a large chamber capable of withstanding the frequent explosions occurring when the deuterium–tritium pellets are heated by the laser beams. The neutrons produced in the nuclear fusion reaction are absorbed in a blanket of sandlike ceramic granules that slide along the inside walls of a doublecone-shaped rotating chamber. The rotation chamber creates a force that holds the granules in place. (If you quickly rotate a bucket of water in a vertical plane, the water does not spill out during the rotation.) When the granules are hot enough they are ejected from the chamber into a heat exchanger where their energy is transferred to a high-pressure helium gas. The heat removed in the heat exchanger is used to produce steam for the operation of a turbine generator in the usual way.

The design of a laser fusion reactor is more flexible than for magnetic fusion reactors. Therefore, a large number of chamber designs can be made. Another famous concept for a fusion power plant was suggested by the scientists of Livermore in the USA and it was given the name HYLIFE (High-Yield Lithium Injection Fusion Energy). The HYLIFE design (see figure 5.16) selects the deuterium–tritium fuel cycle with a liquid-metal blanket. The conversion to electricity is done with a conventional steam cycle. The blanket is made up of a close-packed array of lithium jets. The gaps between the jets allow passage of the high-pressure plasma and they allow for expansion because of the high pressures created in the lithium by the absorption of neutron energy. The X-rays and debris of the pellet hit the lithium streaming jets and vaporize a layer of lithium which recondenses before the next pulse. The neutron flux is attenuated so that the first wall is shielded from radiation damage. The structural wall is made of common ferritic steel

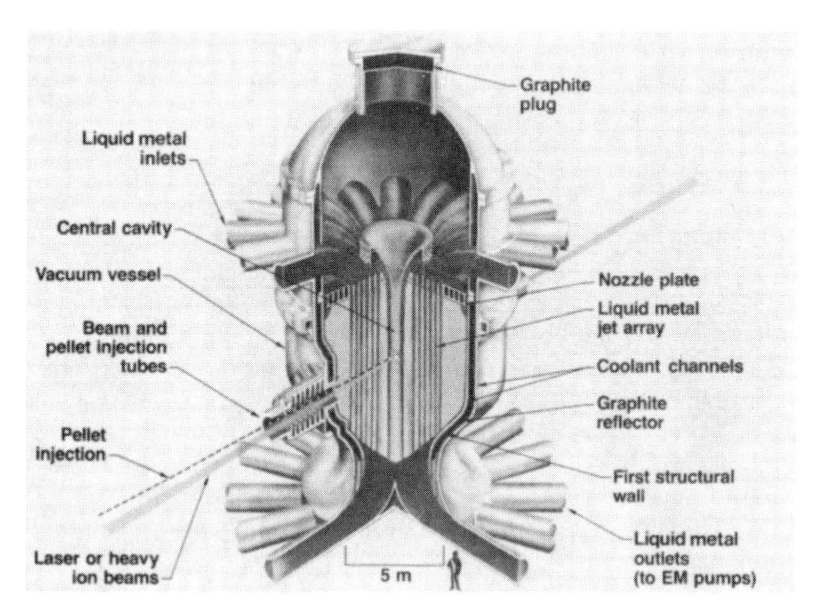

Figure 5.16 A reaction chamber (HYLIFE) in an inertial confinement fusion power plant. (Courtesy of Lawrence Livermore National Laboratory, California, USA.)

and lasts for the life of the power plant. The flowing lithium also regenerates the tritium required for the deuterium–tritium reactions. HYLIFE is an ingenious and an attractive solution to the problems involved in constructing nuclear laser fusion power plants. However, there are still some major technological issues to be investigated which are not yet completely understood.

To summarize, the landmarks in the development of laser fusion research are as follows. Shortly after the discovery of the laser capability of concentrating significant energy and large power into a small space, several scientists recognized that this could lead to the heating of hydrogen isotopes to nuclear fusion temperatures. In 1968, the Lebedev Laboratory of Moscow, in the Soviet Union, headed by the Academician Nikolai Basov (who shared the 1964 Nobel prize in physics for the invention of the laser), was the first to demonstrate that fusion reactions could be produced by focusing a neodymium glass laser beam on a lithium deuteride solid target. The Commissariat à l'Energie Atomique (CEA) Laboratory at Limeil in France reported the production of nuclear fusion neutrons in 1969. Neutron production has been observed since then in all the large laboratories around the world working on the laser fusion program. After the declassification of the spherical compression concept in 1972 by John Nuckolls and his colleagues from the Lawrence Radiation Laboratory at Livermore in California, large laboratories were built in the USA, Japan, the Soviet Union, Germany and China.

For novel ideas declassified and developed during the last decade, see Sections 5.20 and 5.21.

5.16 Particle Beam Fusion

Inertial confinement fusion is based on the principle of heating a small spherical pellet and squeezing it to very large compressions. The pressure that compresses the center of the pellet, where the nuclear fusion fuel is placed, is exerted by the material surrounding the thermonuclear fuel. It was explained previously how this goal can be achieved with powerful laser beams. During the early 1970s it was realized that relativisitic electron beam accelerators can drive inertial confinement fusion implosions. This idea was independently recognized and developed at the Kurchatov Laboratory in the Soviet Union and at Sandia Laboratory in Albuquerque, New Mexico in the USA. After a few years of research, it was realized that electron beams are not appropriate for compressing pellets to high densities. Electrons can penetrate a small pellet as they can move long distances before they are stopped, and thus they can heat the pellet but not compress it. However, ion beams are better candidates for compressing and igniting small pellets because the range of ions (the region within which energy is deposited) is much shorter than that of electrons (with equal energy). It was also realized technologically that relativistic electron accelerators can also be used to accelerate light ions efficiently. The main research that led to this development was done at Cornell University in New York and at the Naval Research Laboratory in Washington, DC. The Sandia Laboratory has modified its electron drivers to produce light ions for nuclear fusion processes. This change in Sandia's program was made during 1979. Lithium ion beams with huge currents of millions of amperes per square centimeter are produced for this fusion program.

The physics of energy absorption and transport seems to be more favorable than for laser-induced inertial confinement fusion. However, the problem of focusing intense ion beams onto a small pellet seems to be more difficult than the focusing of laser beams. Ions have positive charges and therefore the electric (Coulomb) interactions have a tendency to disperse the ion beam rather than to converge it onto a small focal spot. Lasers are composed of photons which do not have any charge, and therefore it is much easier to focus and concentrate energy on small volumes with lasers than with ion beams. If one looks at the overall problem of inertial confinement, it appears that light ion fusion is more favorable than the laser fusion of small pellets. A photon does not have any mass, and consequently it is harder to compress a target with photons than to compress it with 'heavy' particles like ions. From this point of view, it looks even better to compress the pellet with heavy ions, such as uranium ions or other heavy, ionized atoms (from the top end of Mendeleyev's Periodic Table). Heavy-ion fusion programs started during the mid-1980s at Darmstadt in West Germany and at the University of Berkeley in California, USA. However, it seems that the uniform irradiation of a pellet with ion beams is a 'mission impossible'. See Sections 5.20 and 5.22 for novel possibilities.

5.17 Advantages of Nuclear Fusion Energy

Nuclear fusion reactors can operate only in a plasma state of matter whereby the necessary conditions can be achieved. The major raw material for nuclear fusion is deuterium, which may readily be extracted from ordinary water. This raw material is available to all nations without practical limits. The ocean waters of the Earth contain an effectively inexhaustible supply of deuterium. The second generation of fusion reactors will use only deuterium as the thermonuclear burning fuel. For the first generation of fusion reactors one expects a deuterium–tritium mixture to be used. The required tritium will be produced from lithium, which is available from land deposits or from seawater. The lithium supply for this purpose can last for thousands of years, so that the fuel supply for nuclear fusion does not pose any problem. The amount of deuterium and tritium in the reaction area is so small that a large uncontrolled release of energy is impossible. The whole idea of inertial-controlled thermonuclear fusion energy is based on very small quantities of explosion at each time. Any malfunction of the magnetic confinement devices will cause the hot plasma to run into the walls of its containment vessel and cool, so that the fusion reactions will stop at once. No nuclear accident is possible in a nuclear fusion power plant.

In thermonuclear reactors, no fossil fuel is used so that there is no 'dirty' radioactive waste, and therefore no disposal problems. In the first-generation reactor schemes radioactivity might be produced in small quantities by the neutrons which are released during the deuterium-tritium fusion. When the neutrons hit some materials they can produce radioactive elements. However, proper selection of materials is expected to exclude this problem. In such a reactor there is no release even of chemical combustion products because they are not produced. Nuclear fusion is ecologically the cleanest possible source of energy.

Another significant advantage is the fact that the materials and the by-products of nuclear fusion are not suitable for use in the production of nuclear weapons. Although the project of controlled thermonuclear fusion was initiated from weapons programs, the end result will be the development of energy resources for our civilization without any implication for weapons development and construction.

One has to remember and to understand that no civilization is possible without energy. Energy is vital for the survival of the human race. The only source of large quantities of energy for civilizations to come is nuclear fusion energy, the same source of energy as found in our Sun and the stars in our Universe. Plasma physicists are able to build a small sun in the laboratory today and hopefully in a nuclear reactor in the future. A greater international effort should be put into achieving this goal so that by the end of the twenty-first century New York will not become the morbid picture portrayed in the science-fiction movie *Soylent Green.*

5.18 The Transition to the Fusion Era

There is no doubt that the most important issue in plasma physics today is the drive to achieve controlled thermonuclear fusion. This source of energy will be accessible in perpetuity to mankind. This goal is the primary stimulus in the development of laboratory plasma physics.

The American physicist Marshall N. Rosenbluth, who is considered to be the leading plasma theorist of this generation, summarized very clearly the past, the present and the future of magnetic thermonuclear plasma research when he was presented the Fermi award in physics in 1986.

'Why, after thirty years of effort, am I so optimistic about fusion's prospects for eventual success? ... theory and experiment are converging, and within the parameters of our understanding lie the parameters for success.... The 1950s were a decade of innocence when we hoped to succeed without really understanding what we were doing.... The 1960s were the decade in which a fundamental framework of theory was laid down.... The 1970s saw a great increase in support for fusion research due in part to renewed hopes raised by Tokamak results and in part to the oil crisis of those days. With these new funds the needed technology could be developed and smarter, often costlier, but more successful experiments deployed.... Now in the 1980s the world's break-even size Tokamaks are on line.... The exploration of our frontiers will continue. We now stand in fusion research on the threshold of scientific success.'

The beginning of controlled thermonuclear fusion is best defined by the famous physicist Edward Teller in his introduction to the book *Fusion*.

'As soon as the first successful thermonuclear device was exploded discussions on controlled fusion became of great interest in the United States. To administrators and politicians alike it appeared that what was possible in a violent reaction should be put to work in a controlled and peaceful manner. It was necessary, though barely possible, to convince decision makers that achieving the controlled process is incomparably more difficult and that a considerable period of research, experimentation and development must precede any practical application.

Most fortunately, strong support for research was accepted as a first step ...'

Independently of the US program, similar projects began simultaneously in the Soviet Union and in the United Kingdom in the early 1950s. It is not surprising that these projects were initiated after the explosion of the thermonuclear device known to the public as the hydrogen bomb.

More than 20 years before the thermonuclear fusion bomb was exploded, a member of the Soviet Union Politburo, Nikolai Bukharin, approached a young physicist by the name of George Gamow, after having heard his lecture on the fusion energy of the Sun. Bukharin offered Gamow the entire electrical power of the city of Leningrad for one hour each night to carry out research which would lead to fusion energy production. Gamow turned down this generous offer since scientifically the time was not ripe for such an elaborate scheme. Gamow later emigrated to the United States and became a renowned physicist. Bukharin was executed by Stalin after a show trial in 1938 and was rehabilitated by the Gorbachev regime in 1988. It is interesting to note here that Bukharin, a politician, had the vision to try to reproduce on Earth what is happening in the Sun; today, this is the main essence of the controlled thermonuclear fusion program.

Controlled thermonuclear fusion is based on very inexpensive and inexhaustible fuel: the deuterium which is contained in water. From the environmental point of view, nuclear fusion is inherently a safe system. Radioactive fissionable materials are not involved in the fusion process and radioactive waste disposal does not exist. The thermonuclear reactor is safer and cleaner than any other electrical power plant of today. These advantages were recognized and controlled thermonuclear fusion has become a major international effort. Japan has declared fusion to be a national goal. There are large fusion programs in Western Europe, the Soviet Union and the United States, while smaller countries are following this project with smaller plasma programs and through a very fruitful international collaboration. Controlled thermonuclear fusion requires extremely high temperature, as high as in the center of the Sun or even higher. This plasma is confined by very strong magnetic fields in different possible configurations known as magnetic bottles.

An alternative approach to reach the desired conditions for thermonuclear fusion is to compress small pellets to extremely high densities by using high-powered lasers or ion beams. In this case the fuel ignites and burns before the compressed pellet disassembles.

The history of inertial confinement fusion is relatively new and only began in the late 1960s. The main short-term applications were intended for military purposes. The American physicist who was in charge of building the huge laser facility at Livermore, John L. Emmett, stated in the introduction to the 1981 Laser Program Annual Report of the Lawrence Livermore National Laboratory: 'The goal of the ICF (Inertial Confinement Fusion) is to produce significant well-diagnosed thermonuclear experiments with laboratory laser facilities. In the near term, these experiments are for military applications (for example understanding material responses under very high temperatures and pressures or studying processes that occur under extreme conditions experienced in nuclear weapons). With high gain targets, simulations of nuclear explosion effects can also be obtained.'

The long-term goal of ICF is to acquire energy production. The combination of small pellets with a significantly larger energy than invested originally (the scientists call it high-gain targets) driven by an efficient laser or ion beam system to compress the pellets can make available the ultimate fusion power for energy purposes.

At the moment it appears that the magnetic confinement approach is more mature. The inertial confinement approach has the desirable feature that the driver facility (a laser system or an ion beam system) and the chamber of the nuclear reactor can be separated from one another. It is easier to design a nuclear reactor for this approach to fusion. Generally speaking, inertial confinement fusion is a complementary program to the magnetic confinement schemes. Both approaches should be pursued in the future. It is conceivable that large 'magnetic bottles' and very tiny 'pellets' will both be successful in achieving commercially controlled thermonuclear fusion energy. If scientists can sustain the present rate of progress, and, even more important, if governments will decide to give priority to these projects, then it is believed that about three to four decades from now fusion power reactors will begin supplying some of the world's energy needs.

In order to get a 'feeling' for the advancement toward the above 'dream', a short summary of the development of nuclear fusion in the 1990s will be given in the next three sections.

5.19 TFTR, JET and other Magnetic Fusion Devices

The Tokamak Fusion Test Reactor (TFTR) in Princeton, USA, one of the largest Tokamaks in the world, was originally designed to use deuterium fuel in order to advance the magnetic fusion research. Although this device had no prior system to handle tritium (a radioactive element) continuously, the US Department of Energy decided to 'burn' deuterium–tritium before shutting down this facility. In 1996 the TFTR successfully burned deuterium–tritium and achieved a plasma fusion power of 10.7 megawatt (MW) during a time period of 0.4 seconds. This was a big scientific achievement and a giant step forward in the nuclear fusion research before its closing in 1997.

The Joint European Torus (JET) at Abingdon, near Oxford in the UK, was originally designed to use deuterium–tritium (D–T) with special care and remote handling systems. In 1997 the D–T experiments at JET set a record of fusion power of 16.1 MW during a time period of 0.85 seconds and a ratio of fusion power to plasma input power of 0.62, defined as the 'gain'. This scientific 'break-even' was a historical event in the advancement of magnetic confinement fusion. In these experiments it was also observed that the ^{4}He (created in the fusion reaction of D–T) particles are significantly heating the D–T plasma, another important step towards a fusion reactor. It is still a long way towards an economic fusion reactor, requiring a gain larger than 30 and an output power of about 1000 MW in continuous operation.

In Japan, the Atomic Energy Research Institute at Naka-Machi has operated the JT-60U Tokamak since 1991 and very impressive results have been obtained in the product of 'plasma density $\times$ plasma confinement time $\times$ plasma temperature'. This product is related to the famous Lawson criterion for achieving gain in a nuclear fusion reactor.

The General Atomic Tokamak in San Diego, California, named DIII-D, has concentrated on research aimed to enhance the commercial attractiveness of the Tokamak as an energy producing system.

Innovative research is done at the Massachusetts Institute of Technology (MIT) in Cambridge, USA, with a compact high magnetic field Tokamak called Alcator C-Mod.

A complementary range of small to medium Tokamaks are operating in many countries around the globe. These devices are concentrating mainly on plasma physics research generating the database for the extrapolation into the plasma domain for plasma fusion reactors.

Many non-Tokamak devices are also being developed in different countries, checking their capabilities for a fusion nuclear reactor. Most of the non-Tokamaks are toroidal (doughnut); the difference is their configuration of the magnetic fields. While alternate magnetic concepts may differ from Tokamaks in geometry size, input power and technology, the objective of heating, confining and burning a D–T plasma to yield a net energy gain is unchanged.

The stellarator, which unlike the Tokamak does not carry an externally induced current in its plasma, also has a toroidal (doughnut) geometry. Confining the plasma are magnetic fields, which are generated by external helical currents (of opposite directions in alternate coils) and toroidal currents. The magnetic fields induced by these currents have the direction of a spiral along the torus. Whereas in the Tokamak one of the magnetic fields confining the plasma is created by the plasma current, in the stellarator all the magnetic fields confining the plasma are generated only by external currents. The fact that all the confining currents are external is an advantage, since in this way the continuous steady state operation is much easier. Unlike the pulsed Tokamak, the stellarator works in a steady state. However, geometrical simplicity and axial symmetry are lost in the stellarator in comparison with the Tokamak.

Another toroidal confinement device which has received great attention is the Reversed Field Pinch (RFP). This device is like a Tokamak, but the toroidal magnetic fields change direction near the plasma boundary, hence the name Reversed Field. This reversed field is generated naturally by the plasma in a process known as the dynamo effect. The plasma current and therefore the poloidal magnetic field is much stronger than in comparable Tokamaks. In an RFP plasma the poloidal and toroidal magnetic fields are about equal, while in a Tokamak the toroidal magnetic field is much larger than the poloidal magnetic field. Here it is important to remember that the poloidal magnetic field is produced by the strong plasma current flowing around the torus. The RFP device seem to be more stable than Tokamaks as far as magnetic hydrodynamic instability is concerned.

Although so far these devices are less developed than the Tokamak, some scientists believe that the RFP or the stellarator may be more compact, cheaper and more stable and perhaps the answer to the future fusion reactor.

In 1998 a large helical device (LHD) successfully began to operate at the National Institute for Fusion Science at Toki, Japan. This is a large step toward a stellarator power plant. So far the other non-Tokamak devices are small in comparison with existing large Tokamaks (JET).

5.20 Indirect Drive for Inertial Fusion Energy

For direct drive inertial confinement fusion, 1% laser uniformity of irradiation of the pellet is required in order to avoid disastrous hydrodynamic instabilities. To overcome this an indirect drive scheme was suggested by scientists at the Lawrence Livermore National Laboratory in the USA. Since this idea might resemble the concept of a hydrogen bomb, the indirect drive research was classified by countries possessing thermonuclear weapons. The declassification of this subject by the USA in the 1990s has led to the advancement and international collaboration in inertial confinement fusion in general.

In the indirect drive scheme the laser beams hits the internal envelope of a cylinder (called hohlraum—German for 'cavity' or 'hollow'). This cylinder, a few centimeters long, is composed of a material with a high atomic number (called high Z) such as gold (see figure 5.17). The atomic number of gold is 79, that is, the gold atom has 79 protons and 79 electrons. It was found that most of the laser energy is converted (at the cylindrical vessel) into soft X-rays with a radiation temperature of up to a few hundreds of electron volts. One can see the cylinder as a fuel cavity containing soft X-ray photons. The center of the cylinder contains a pellet with D–T fuel, as in the case of the direct drive. This pellet is now uniformly irradiated by the soft X-rays. If everything goes well, the pellet absorbs the X-rays, is compressed and ignited. In this way the uniformity of irradiation is significantly improved in comparison with the direct drive. However, since the indirect scheme first converts the laser energy into X-rays and only some of these X-rays hit the pellet, the energy absorption is less in comparison with the direct drive.

The indirect drive is applicable also with ion beams (see figure 5.17). Since it seems almost impossible to irradiate a small pellet uniformly with ions, the indirect drive scheme seems to be the only way to obtain fusion energy with ion beams. In this case the ion beams are focused on to a high-Z target at the two bases of the cylinder, are then absorbed, their energy is converted into X-rays and deposited inside the cylinder where an X-ray cavity is created. As in the previous case, the D–T pellet at the center of the cylinder is compressed and heated to nuclear fusion conditions.

A major possible advantage of the indirect drive as compared with the direct drive is the reduced instabilities during pellet compression. This

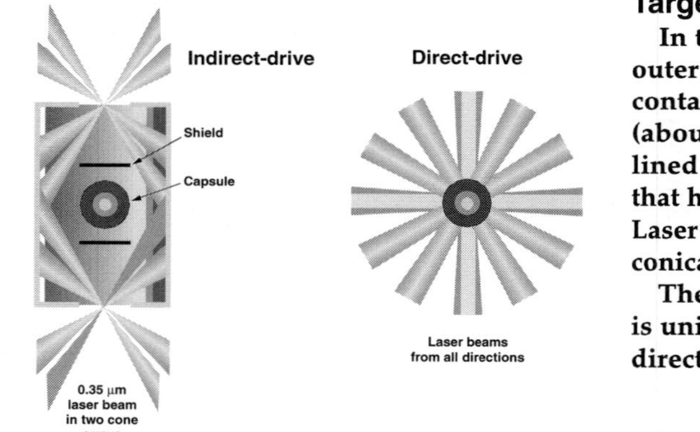

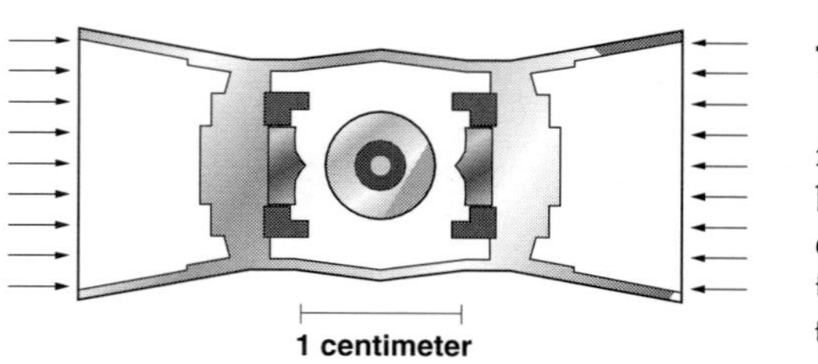

Figure 5.17 Targets for solid-state laser and heavy-ion driver. (Courtesy of the University of California, Lawrence Livermore National Laboratory and the US Department of Energy.)

feature, together with the demonstration of uniform energy deposition over an entire pellet surface, is very promising for future advancements. However, due to the reduced energy absorption by the pellet and the increased complexity of hohlraum manufacture the competition is still open between the direct drive and the indirect drive schemes for inertial fusion energy.

5.21 Fast Ignitors

In the conventional direct or indirect inertial confinement fusion the following steps are necessary in order to get high gain:

1. The lasers (direct drive) or the X-rays (indirect drive) irradiate uniformly the fusion target (a pellet with D–T). The absorbed energy heats the surface of the pellet and a plasma 'atmosphere' is created.
2. The plasma expands outwards from the pellet and, as in a rocket, the pellet is accelerated toward its center (implodes) and compresses the fuel.
3. The irradiation profile and target are designed in such a way that during the final stage of the implosion, a hot central core surrounded by a cold very dense D–T fuel is created. A spark of fusion is created at the center of the fuel.
4. Thermonuclear burn spreads from the center outwards until most of the D–T fuel is exhausted.

It is very difficult to design the fusion target to allow steps 3 and 4 to occur. Moreover, due to hydrodynamic instabilities it is uncertain whether step 3 can be achieved. Even if step 3 is achieved would step 4 be a stable process?

In order to overcome these problems, in 1992 Max Tabak with his colleagues at the Livermore National Laboratory in the USA suggested a new scheme, named 'fast ignitors'. In this scheme steps 1 and 2 are identical to the conventional schemes. At the moment of maximum compression (more or less uniform compression) a very short high-intensity pulse ignites a corner of the fuel. The thermonuclear burn spreads very quickly from that corner throughout the compressed fuel. Thus the difficulties in steps 3 and 4 are overcome.

This scheme might be possible with the development of extremely 'very short' and 'very high power' lasers, the so-called 'femtosecond lasers' (1 femtosecond (fs) is equal to one part of one million of billions of one second). Today there are very high power lasers with pulse duration from as short as 10 fs to as 'long' as 1000 fs. For the fast ignitor scheme extremely high power lasers with a duration of about 10 000 fs are needed. The pressure of this laser light is so high that it 'pushes' the plasma atmosphere inside allowing the laser to interact directly with the high density target. In the normal case the laser is stopped outside in the plasma atmosphere far from the solid density. Moreover, due to the very high power of the laser beam, very energetic electrons can be created to penetrate the compressed fuel and heat part of it to large enough temperatures necessary for ignition.

This scheme of 'fast ignitor', like the other schemes of conventional direct or indirect drive, have not been proven experimentally so far. The ignition and the thermonuclear burn of a small target (few millimeters in diameter) is still awaiting the development of larger and more intense laser drivers. So far thermonuclear ignition has been achieved only in the hydrogen bomb.

5.22 The Z-Pinch

The research on Z-pinch goes back to the beginning of magnetic confinement in the 1950s. What is the original idea of the Z-pinch? A very strong electric current is created through a deuterium (or deuterium–tritium) gas along a direction labeled as the Z-axis. This high current ionizes the gas and generates a strong magnetic field surrounding the current which pinches the plasma. The compressed plasma is heated to high temperatures and fusion was expected many years ago. So what went wrong? It turns out that this Z-pinch process is not stable, flow instabilities break the plasma into blobs and the plasma cannot be compressed and heated as necessary for fusion reactions.

Although fusion was not successfully achieved with Z-pinches, X-rays were successfully generated. For the past 30 years Z-pinch devices have been developed in the USSR, US and UK in order to create powerful X-ray sources. A key issue for these devices is to create very high power and large electrical currents during a short period of time. The technology used for this purpose is called 'pulsed power'.

In a pulsed power system the energy is stored in capacitors and special design electrical switches discharge this energy in a short period of time. For shorter discharge times one gets higher electrical power. The pulsed power devices are very efficient in converting the stored energy into useful energy for the generation of the high current of the Z-pinch device. In 1997, a Z machine at Sandia National Laboratory in Albuquerque, New Mexico, US, has reached about 300 terawatts (one tera is equal to million millions) generating two million joules of X-rays in a few nanoseconds. The X-rays created in a Z-pinch device like that of Sandia can be used for indirect drive scheme as described in the previous section.

The great breakthrough with the Z-pinch devices came in 1995 when scientists at Sandia used a wire array of tungsten to form a cylindrical shell. During the electrical discharge a high current passes through these wires causing them to explode and create a uniform plasma in the shape of a cylinder shell. This shell of plasma implodes on to a filament positioned at the center of the shell. During the collision between the 'plasma-wires' and the filament a big burst of X-rays occur. The scientists at Sandia believe that an increase in their Z-pinch device by a factor of three will produce enough X-rays to induce ignition indirectly into a fusion pellet containing deuterium and tritium.

5.23 Outlook

The present resources of our planet Earth are limited. A new source of energy is desperately needed if our civilization is to survive. Scientists today believe that there is a way of producing energy which can last in perpetuity. We have learned in this book that understanding plasma—the fourth state of matter—is crucial in achieving this goal. Controlled thermonuclear fusion is ecologically the cleanest possible source of achieving large quantities of energy. The major raw material for these schemes, deuterium, is available to all nations without any limits. Furthermore, as most of our Universe, including our own Sun, is in a plasma state of matter, it is surprising that intellectuals and the public at large are not aware of plasma physics.

We ask ourselves what is the reason for this ignorance? We believe that this subject is 'inscrutable' and thus taught very little at university and not at all in high schools. Plasma physics is a subject which cannot be understood without a preliminary acquaintance with many disciplines of physics. These include an integration of different subjects such as atomic physics, nuclear physics, electromagnetism, fluid mechanics, statistical mechanics and quantum mechanics. In some sense it is the integration of most of the branches of physics and it is therefore difficult to teach and even harder to comprehend. However, due to the intellectual and practical importance of this field, we feel that a major effort must be made in order to bring this important subject to the knowledge of our students in particular, and to the public at large in general. The children of today will be the important decision makers of tomorrow; the future politicians, scientists, members of the community. They will be living in a future based on nuclear fusion energy: a future where plasma physics and technology play a crucial role in the development of their society. We feel that, in order for them to make the right decision intelligently, we of today are obliged to teach and introduce the plasma state of matter to those of tomorrow. By 'invading the inscrutable' we believe that we have paved the way for the understanding of the fourth state of matter.

Chapter 6

...More History of Plasma Physics

An extension of the history of plasma physics as described throughout the book

6.1 Plasma Without Realization

With the development of the science of electricity in the 19th century, scientists created, without realization, plasma effects. In the 1830s the famous English scientist Michael Faraday played with electrical discharges which exhibited strange glows unlike in any known state of matter (solids, liquids or gases). The understanding of these strange phenomena was not possible before the discovery of the electron by the English physicist Joseph John Thomson in 1897 and the measurement of the electron charge in 1911 by the American physicist Robert Andrews Millikan. Moreover, understanding of the atomic theory was crucial before the fourth state of matter could seriously be considered. The atomic model was first suggested in 1913 by the Danish physicist Niels Bohr and it was confirmed experimentally in 1920 by the New Zealand-born English physicist Ernest Rutherford who carried out his experiments in England.

The understanding of atomic physics is not the story of only one or two great men. It is the story of many great scientists working together during a very exciting era of great collaboration between theoretical and experimental physics. The American scientist Robert Oppenheimer, the head of the Manhattan Project (the development of the first atomic bomb in Los Alamos, USA), a great physicist and intellect, summed up this era the following way. 'Our understanding of atomic physics has its origins at the turn of the 19th century and its great synthesis and resolutions in the nineteen-twenties. It was a heroic time. It was not the doing of any one man; it involved the collaboration of scores of scientists from many different lands, though from first to last the deeply creative and subtle critical spirit of Niels Bohr guided, restrained, deepened, and finally transmuted the enterprise.'

6.2 Realizing the Fourth State of Matter—Plasma

In 1879, the English physicist Sir William Crookes, while considering the properties of matter in electrical discharges, suggested that these gases are the fourth state of matter.

The terminology 'fourth state of matter' follows from the idea that by heating a solid sufficiently, one can usually obtain a new state of matter—a liquid. If one heats the liquid further, it undergoes a phase transition to a gas. The addition of more heat results in the ionization of some of the atoms. The behavior of the last phase, namely of the ionized gas, is distinctly different from that of regular gases. Crookes already realized in 1879 that the matter of a discharged gas behaved unlike a regular gas. The interactions between the particles of the plasma are dominated by the Coulomb force. This force is named after the French physicist Charles Augustin de Coulomb who in 1785 measured the force between magnetic poles at varying distances. The same law of force was later found by Coulomb to describe the interaction between two electrically charged bodies. The fourth state of matter exhibits a collective behavior because of the long-range nature of the Coulomb force between charged particles. For example, a positive ion can feel the force of attraction from many electrons surrounding it. All the electrons in a sphere of a specific radius surrounding the ion will influence the motion of the ion. The radius of this sphere was named after the Dutch physicist Peter Debye and it is called the Debye radius or length. Similarly, an electron will be influenced by the other surrounding charged particles. Therefore, the motion is determined by collective interactions and not only by the force between two individual particles. John William Strutt Rayleigh, already known as Lord Rayleigh in 1906, described for the first time this collective behavior in his analysis of electron oscillations inside the atom.

The physics of plasma is the story of *electrons* and *ions*. Both are charged particles; the electron is negative while the ion can be either negative or positive. The story of ions, like the story of electrons, started long before the atomic picture was unveiled at the beginning of the 20th century. In 1832 the famous English scientist Michael Faraday developed the laws of electrolysis. He described the formation of electrically charged particles in solutions in which two rods were installed and connected to a battery. Solutions of most inorganic acids, bases and salts conduct electricity and are therefore called electrolytes. The electrically charged particles in these solutions are positive and negative ions. For example, when sodium combines with chlorine to form sodium chloride (common salt), each sodium atom transfers a negative charge (today we know that this is an electron) to a chlorine atom, thus forming a positively charged ion of sodium and a negative ion of chlorine. In 1887 the Swedish chemist Svante Arrhenius was the first to realize that substances in solution are in the form of ions and not molecules, even without applying an electrical potential (through a battery for example). Arrhenius stated that for a given quantity of sodium chloride, one will get a larger amount of ions if the solution contains a larger quantity of water. The theory of the dissociation of electrolytes was further developed by the famous Dutch physicist Peter Debye in 1923. Debye's theory, which is the accepted theory of solutions, assumes that the electrolytes (e.g. sodium and chlorine) are completely dissociated in solutions. The tendency of the ions to migrate and conduct electricity depends on the concentrations of the electrolytes and on the properties of the solvent such as the dielectric constant. The dielectric constant describes the ability to store electrical charges and consequently electrical energies when placed between parallel plates (a capacitor) connected to a battery. For larger dielectric constants larger quantities of electrical charges and energies can be stored. The ionization phenomenon is most prominent when water is the solvent, as it has a high dielectric constant.

Debye's theory of ions in solutions is sometimes very appropriate for describing the ionization state in gases. The physics of electrolytes in solutions, as well as the physics of electrons and ions in electrical discharges, is based on the properties of long-range Coulomb forces.

The American scientist Irving Langmuir introduced in 1928 the word *plasma* to explain some phenomena occurring in these electrical discharges. He and Levi Tonks, another American physicist, found that during an electrical discharge there are regions of plasma where the electrons oscillate in a collective manner. The motion of the electrons inside the plasma was described by Langmuir and Tonks by using simple laws of fluid motion (remember the conservation laws in Chapter 2) together with the laws of electrical forces (Coulomb forces).

In 1946, the Soviet physicist Lev Davidovich Landau, one of the greatest scientists of the 20th century, developed the theory which describes the interaction between particles and plasma waves. He pointed out that a particle can gain or lose energy by colliding with a wave, even without direct collisions with other particles. Landau's paper is crucial in understanding the nature of hot plasma in stars, as well as in laboratory experiments. Modern plasma physics starts with Landau's remarkable contribution.

Although about 99% of our Universe exists in a plasma state, plasma does not exist naturally on our planet Earth. One of the exceptions for natural plasma creation is during lightning in a thunderstorm. In some sense the lightning in nature is similar to a large electrical discharge in a laboratory. The reason that plasma is so scarce here on Earth is that plasma is created at very high temperatures while our weather is too cold to produce plasma. Moreover, creation of plasma by radiation from the Sun is excluded on Earth by our atmosphere which absorbs the radiation that can produce a plasma. Plasma on Earth is produced in laboratories by heating gases to high temperature or by the electrical discharges described by Crookes, Langmuir, Tonks and their followers.

6.3 Controlled Lightning

Benjamin Franklin, the famous American scientist and statesman, began his electrical experiments around the year 1747. Using his well-known kite experiments during thunderstorms, he identified the lightning phenomenon with electricity. His invention of the lightning rod that neutralizes the accumulated static electricity is still used today to save human lives and to avoid damage to property. Benjamin Franklin can be referred to as the first electrical engineer with the invention of his rod.

In 1808, the British scientist, Sir Humphrey Davy, discovered the DC arc discharge. Davy's self-educated assistant, who became one of the greatest experimentalists of all times, Michael Faraday, developed the high voltage DC electrical discharge tube in 1830. These electrical discharges act like a controlled thunderstorm in a laboratory. Some of the technical scientific terms introduced into our vocabulary were contributed by Faraday: the electrode (in Greek electrical road), anode (in Greek road up), cathode (in Greek road down) and ion (in Greek to go).

The characteristics of low pressure DC electrical discharges were discovered by English and German physicists of the nineteenth century. During an electric discharge cathode rays flow from the cathode to the anode. In 1860, the German scientist Julius Plucker detected a fluorescence phenomenon when the cathode rays hit some non-metallic substances found in ores such as sulfur. Plucker also discovered that cathode rays are diverted from their trajectory by magnetic fields. This effect became vital during the development of modern electronic devices such as the television.

Very important research in discharge phenomena was carried out by the English scientist Sir William Crookes who also introduced the term 'the Fourth State of Matter'. Around 1880, he developed a very useful vacuum tube (Crookes's tube) which enabled him to investigate the conduction of electricity in plasmas. These simple experiments led to the conclusion that the Fourth State of Matter is a very good conductor.

In 1883, Thomas Edison, one of the greatest inventors of the nineteenth century, while experimenting with vacuum tubes, accidentally discovered that when heating one of the electrodes, a flow of current appeared between the hot and cold electrodes. This was later referred as the 'Edison effect' and laid the foundation for the electronic industry.

Nikola Tesla, another great scientist and inventor of the end of the nineteenth century, discovered the alternating current dynamo, transformers and electric motors: practically the basis of all alternating current machinery. After migrating to the US, he worked with Edison. Later on he opened up his own laboratory where he also produced high-voltage discharges in air. In this way he created a 'controlled thunderstorm'. While conducting the latter experiment, using several million volts, he placed himself inside a Faraday Cage (an area surrounded either by metal rods or by metal netting) reading a book.

In 1906 Sir Joseph John Thomson was awarded the Nobel prize in physics for theoretical and experimental investigation of the passage of electricity through plasmas in tube discharge. From his investigations, Thomson proved that the cathode rays are negative charged particles. It was the famous scientist, Hendrik Antoon Lorentz, who suggested in 1906 that J. J. Thomson's charged particle be called an electron. Lorentz, who received the Nobel prize in physics in 1902, also explained how electrical and magnetic forces act on electrons. His equations were the first step to describe the motion of the electrons in a plasma.

One of the most important technical breakthroughs came when the German engineer Wolfgang Gaede introduced the rotary mercury pump in 1905 and the mercury diffusion pump in 1911. These vacuum pumps were faster and better, significantly reaching lower densities in a gas tube. This enabled plasma discharge science to push forward.

In 1925, Irving Langmuir advanced the understanding of electrical discharges by explaining the sheath phenomenon. This large potential drop occurs near the electrodes inside the vacuum tube. Langmuir was also the first to coin the word plasma for the fourth state of matter.

Other contributors to the industry of plasma discharges are the British and American scientists who developed radar during World War II. Following the war, the same techniques were used to create microwave discharges in gases to create plasma, which further led to new channels for the applications of plasma in industry.

A major impact on plasma physics was the controlled fusion research which began in the 1950s and is still carried out to the present day. It is a well known fact that while investigating one field many interesting new insights are unfolded in other areas of research. The development of fusion research paved the way to a deeper understanding of plasma physics. As numerical simulations were developed and new diagnostic techniques were introduced, plasma processing in industry gained momentum. The better understanding of controlled lightning has led to the development since 1970 of the microelectronics industry for the production of chips. The application of similar plasma for the processing of industrial wastes has been put to use in the 1990s. It is expected that 'controlled lightning' will be a cornerstone in the high-tech society of the twenty-first century.

6.4 The Ionosphere – A Plasma Mirror for Radio Signals

The existence of a greatly extended conducting layer of plasma in the Earth's upper atmosphere was first suggested in 1882 by the Scottish physicist Balfour Stewart. He proposed the existence of a conductive gas layer in order to explain the variable part of the Earth's magnetic field. The discovery of the important role played by this conducting layer of gas in the upper atmosphere in the transmission of radio signals represents an early chapter in the history of radio. One year after the Italian scientist Guglielmo Marconi received radio signals across the Atlantic Ocean in 1901 and created a world-wide sensation, the existence of the ionospheric layer was suggested in order to explain the radio transmission across the Atlantic Ocean. Marconi's transmission was explained in 1902 by the English physicist Sir Oliver Heaviside and the Indian-born electrical engineer Arthur Edwin Kennelly, who worked in the USA, by assuming the existence of an electrical conductive layer in our atmosphere. This conducting layer of charged particles in the Earth's atmosphere was the same as that already defined by Balfour Stewart in 1882. This ionospheric layer is a plasma state of matter that reflects the radio waves and enables them to follow the Earth's curvature instead of traveling off into space in a straight line.

Electromagnetic waves usually travel in a straight line. The electrons in the plasma reflect the radio waves exactly as the electrons in a mirror reflect visible light. The electromagnetic waves are reflected by electrons if the electron density has the proper value. Every 'color', i.e. frequency of light, is reflected by a particular density of electrons. If the density is smaller than this appropriate number density then the light penetrates the plasma electrons. For larger densities the plasma acts as a mirror. Our ionosphere is a mirror for radio waves. However, shorter wavelengths, such as visible light, can easily penetrate the plasma ionosphere.

This plasma layer in the upper atmosphere is maintained at altitudes of 60 km and higher by solar radiation. Photons arriving from the Sun ionize this layer, i.e. in this process photons are absorbed by atoms of the ionosphere and electrons are removed from these atoms. The experimental demonstration of the existence of this plasma layer, the ionosphere, was conclusively given in 1926 by the American geophysicist Merle Anthony Tuve and by the Russian-born American physicist Gregory Breit.

The English scientist Sir Edward Victor Appleton won the 1947 Nobel Prize in physics for the discovery of the upper region of the ionosphere, the so-called Appleton layer of the ionosphere. The ionosphere layer discovered by Appleton is a 'mirror' for radio waves and for this purpose it is extremely useful in modern communications. Appleton's discovery made possible more reliable long-range radio communication. His research into the propagation of electromagnetic waves also made a major contribution to the development of radar.

6.5 Plasma in Space

In 1942, after the discovery of new solar phenomena, the Swedish physicist Hannes Alfven unified the mutual interactions between ionized gases and magnetic fields by writing the equations describing the motion of electromagnetic fluids. The mathematical solution of this 'new' fluid consists of waves of electrons and ions that were found not only in laboratory plasma experiments but also in the plasma of our atmosphere and Sun. These waves were later called 'Alfven waves'.

Following the first successful launch of an artificial Earth satellite, *Sputnik*, by the Soviet Union in October 1957, modern plasma physics in space began. Satellites launched by the USA and the USSR are able to circle the Earth in about one and a half hours. Many of these satellites are capable of carrying out experiments on space temperature, pressures, magnetic fields and other plasma-related phenomena. In 1960, the solar wind was discovered by a spacecraft. The solar wind is composed mainly of a flux of protons and electrons that are accelerated by the high temperatures of the solar corona (the outer region of the Sun). During the solar flares, there is a significant increase in the solar wind. Today plasma phenomena in space are measured by space stations and interplanetary satellites.

Another plasma discovery in space was accomplished by the American physicist James Alfred Van Allen with the *Explorer 1* satellite in 1958. Van Allen discovered two doughnut-shaped zones of charged particles, such as electrons and protons, which are trapped by the Earth's magnetic field. The charged particles are 'caught' by the magnetic force of our planet Earth and follow corkscrew motions around these magnetic lines of force. The zones, which are most intense over our Equator at an altitude of higher than a few hundred kilometers, are named 'Van Allen radiation belts'. The charged particles in these belts arrive from cosmic rays and from our Sun. Moreover, the high-altitude nuclear explosions carried out by the 'big powers' during the 1950s and the 1960s left a trail of charged particles in the Van Allen belts.

Another plasma phenomenon in our atmosphere is that familiar and beautiful display of light in the sky—the aurora above our North and South Poles. The southern aurora was named by Captain James Cook in 1773 the aurora australis, while observing this phenomenon in the South Indian Ocean. The Australian Aborigines believe that the 'dancing plasma' is nothing else than the dance of the gods while some natives of the southern part of India believe that the aurora is a message from the god Buddha. The northern aurora was named aurora borealis by the French scientist Pierre Gassendi in 1621. This phenomenon of 'dancing plasma' plays an important role in the mythology of the Eskimo, the Scandinavians and other northern nations. The northern aurora was also observed and discussed by the Greek philosophers. The modern investigation of this phenomenon was started in 1784 by the French-born English scientist Henry Cavendish determined the altitude of the aurora phenomenon as between one hundred and several hundreds of kilometers. However, this phenomenon could not be explained before the discovery and understanding of plasma physics.

During the past two decades, a large number of artificial satellites have been sent to traverse the magnetosphere, the region in our atmosphere where magnetic and plasma phenomena occur. These satellites have surveyed and mapped the distribution of the aurora particles and the magnetic fields that 'caught' those charged particles.

The auroral dancing columns of light in a variety of forms are caused by electrical discharges in the upper atmosphere. Around the North and South Poles, one can see different shapes of light formations, such as rays and dancing arcs. This light is emitted from plasmas formed by fast electrons arriving from the Sun and caught near the magnetic poles of our planet Earth. The electrons colliding with the atoms of the atmosphere release more electrons from the atoms and ionization occurs. While the photons from the Sun creating the ionosphere are a permanent phenomenon, the bursts of fast electrons from the Sun that create the aurora are a result of Sun storms and other disturbances. Intense auroral activities may seriously disrupt transpolar short-wave radio communications. This activity is related to electron density changes in regions of our atmosphere and therefore the radio-wave 'mirrors' are affected.

So far we have glimpsed the effects of plasma in our atmosphere. The plasma in space is mainly in the stars. Our Sun is composed of plasma. All the other suns in our Universe (many billions of billions of them) are in the plasma state of matter. We shall not discuss the discoveries of these suns since they are outside the scope of our book. However, in the next section we consider one of plasma's most exciting issues.

6.6 The Sun's 'Secret' Source of Energy

Even before the discovery of nuclear energy, it was clear that our Sun, like other stars in the sky, must contain in its interior some huge unknown energy sources. It was calculated that if our Sun's energy stemmed from a chemical energy source, such as oil, coal, etc, it would have burnt out long ago, and in this case no civilization would be possible on Earth. In the middle of the nineteenth century, the German physicist Hermann L. F. von Helmholtz and later the British physicist Lord Kelvin suggested that the Sun gets its energy supply from a slow but steady contraction (shrinking) of its giant size. In this way the Sun would be able to supply radiation for a period of about 20 million years. However, the geologists required a few billion years to explain the evolution of our planet. Thus the mystery of the source of solar energy remained unsolved.

In 1896, the French physicist Antoine Henri Becquerel discovered radioactivity and this fact revealed the unexpected energy hidden inside the atom. It took another 30 years to develop and understand nuclear physics before the secret of the Sun's energy source was revealed. During this time considerable progress in the understanding of solar interiors was made, mainly due to the work of the British astronomer Sir Arthur Eddington. In 1905 Albert Einstein published the Special Theory of Relativity and his famous equation expressing that mass and energy are equivalent, so that mass can be converted into energy and vice versa.

These developments, together with the epochal experiments of Lord Rutherford on the artificial transformation of elements (the alchemists' long sought-after hope and dream), led the physicists Robert R. Atkinson from England and Fritz Houtermans from Germany to explain the secrets of the energy sources of our Sun. They showed in 1929 that in the central region of the stars, including our Sun, there exist the right conditions of high temperatures and high densities for sustaining thermonuclear fusion reactions. In these reactions two light nuclei, such as hydrogen, for example, fuse together and form another, heavier, nucleus. During this process huge quantities of energy are released. The German-born American physicist Hans Albrecht Bethe was the first to describe accurately in 1938 the nuclear reactions responsible for the production of energy in the stars. The interior of the Sun is in a very hot and very dense plasma state where hydrogen atoms fuse together to form helium atoms. During this thermonuclear fusion a lot of energy is released. The secret of the stars was revealed by nuclear physics inside a plasma state of matter.

Ordinary energy from coal, gasoline or even explosives comes from the fact that electrons in the atoms are disturbed in one way or another. On the other hand, nuclear energy is released by disturbing the nucleus, in particular by splitting a heavyweight nucleus—so-called fission—or by fusing two lightweight nuclei—so-called fusion.

6.7 Splitting the Atom—Fission

Nuclear fission was achieved for the first time by the two German physicists Otto Hahn and Fritz Strassmann in 1938 in collaboration with Lisa Meitner, who had to escape from the Nazi regime and whose name was therefore not included in the famous publication on nuclear fission. Uranium bombarded by neutrons split into two fragments of about equal weight: nuclei of barium and krypton, or strontium and xenon, or any other two appropriate medium-weight nuclei. Neutrons are also released during the fission process, further causing the fission of other uranium nuclei. Let us imagine that when a neutron splits one uranium, two medium-weight nuclei and three neutrons are created. If each neutron again splits a uranium nucleus then we get six new fragments and ($3 \times 3 =$) nine neutrons. These nine neutrons will split nine uranium nuclei releasing ($3 \times 3 \times 3 =$) 27 neutrons, etc. This process is usually called a chain reaction. The actual multiplying factor is less than three and closer to two as some of the neutrons are lost. However, the rate of propagation is still extremely high. During the fission of uranium we gain energy. The weight of the end result of the fission reaction is less than the weight of the uranium nucleus. Therefore, in the fission process, part of the mass is missing. This missing mass is transformed into energy. The atomic explosion caused by the atomic bomb is the result of an uncontrolled chain reaction in a fissionable material such as uranium or plutonium. On the other hand, the energy released in a controlled chain reaction of uranium, for example, can be used for the production of electric power, as is done today in hundreds of nuclear reactors around the world. Nuclear fission was discovered for the first time in December 1938. Controlled fission was accomplished at the end of 1942 under the guidance of the famous Italian-born American physicist Enrico Fermi. The first uncontrolled and violent explosion occurred in the summer of 1945 under the leadership of the American physicist J. Robert Oppenheimer who headed the Manhattan Project at Los Alamos in the USA.

6.8 Fusion – The Synthesis of Light Nuclei

The story of nuclear fusion is quite different. The uncontrolled fusion device known as the hydrogen bomb was exploded by the United States in 1952 and soon afterwards by the United Kingdom and the Soviet Union. The 'father' of the American hydrogen bomb is the distinguished Hungarian-born American physicist Edward Teller. The 'father' of the Soviet hydrogen bomb is the well known physicist Andrei Sakharov, who later won the Nobel Peace Prize for his accomplishments as a Soviet fighter for freedom and peace.

As soon as these thermonuclear devices were exploded, scientists and politicians alike became interested in controlled fusion for energy production. However, to achieve this goal is incomparably more difficult than it is to obtain a violent thermonuclear explosion. A considerable period of further research, experimentation and technological development will be necessary before a controlled nuclear fusion reactor can be achieved.

Thermonuclear fusion is the synthesis of light nuclei into heavier ones, such as the fusion of hydrogen nuclei into a helium nucleus. Nature has accomplished the fusion of hydrogen in stellar interiors, such as in the interior of our own Sun. The Sun gains all its huge amount of energy from thermonuclear processes while synthesizing hydrogen into helium. The temperature in the interior of the Sun needed for the fusion process is enormous—about a hundred million degrees. Such high temperatures are achieved in the hydrogen bomb when triggering it with an atomic bomb explosion. But how can this be accomplished in a controlled manner in the laboratory? How can we build a furnace in a laboratory or in a reactor to sustain temperatures similar to the ones in the interior of our Sun without destruction? The crucial condition for fusion is obtaining and maintaining extremely high temperatures. Even a temperature of a few thousand degrees Celsius is technologically very difficult to sustain and control. Thus, sustaining the many millions of degrees needed for fusion reactions seems an impossible task. The hydrogen bomb, the Sun and the stars achieve the high temperatures violently and by brute force. Therefore, man-made controlled fusion is a formidable task. Research is in progress around the world to develop techniques for releasing controlled fusion energy and converting it into useful electrical power.

There are so far two opposite approaches by which scientists hope to achieve this goal; one approach is to use a dilute gas of hydrogen isotopes and thus to slow down the violent reactions. Moreover, the dilute hydrogen isotopes are kept far away from the furnace wall by strong magnetic fields. This of course can be achieved in a plasma which can be confined by huge magnetic bottles. Magnetic bottle is the name devised by Edward Teller to describe a composite configuration of magnetic forces that is able to contain a hot plasma. A famous example of a so-called magnetic bottle is the large plasma chamber in the shape of a doughnut known as the Tokamak which was invented in the Soviet Union in the 1950s.

The second approach is to bring the hydrogen fuel to extremely high densities, more than a thousand times that of liquid hydrogen. These explosions are similar to those occurring in a hydrogen bomb but each individual explosion is very, very small—more than one million times smaller than a hydrogen bomb—and can thus be contained in a vessel without damaging it. This process is repeated in a controlled way, resulting in nuclear energy production analogous to that of an internal combustion engine. The 'trigger' of this approach is suggested to be very powerful lasers or equivalently very powerful beams of particles. In all these thermonuclear processes, plasma is created. The physics of controlled fusion is today the physics of plasma. The nuclear physics of this process is well known; however, the plasma physics is not yet completely understood and more research and development is needed.

6.9 Solving the Energy Problem for the Generations Ahead

A significant contribution to controlled thermonuclear fusion research was carried out by W. H. Bennett in 1934 and by L. Tonks in 1939 on the plasma pinch phenomenon. A plasma column (line) carrying a large electrical current tends to contract in the radial direction. That is, the plasma is 'pinched' or 'squeezed' due to the interaction between the current in the plasma and the magnetic field that this current creates. (A current creates a magnetic field around itself. Plasmas are good conductors of electricity and therefore large currents can be passed through them.) The plasma 'pinch effect' plays an important role in the development of controlled thermonuclear reactions. However, in 1939, scientists did not yet realize the importance of Bennett's and Tonks's papers for the terrestrial production of thermonuclear reactions in magnetically confined plasma. Work on thermonuclear fusion began at the end of the 1940s in the USA, UK and USSR following the development of the nuclear bomb.

The thermonuclear project using magnetic confinement devices known as magnetic bottles began in the United States in 1951 as a classified program known as Project Sherwood. Similar classified programs began about the same time in England, in the Soviet Union and in France. It soon became very evident that the physics of magnetic bottles is very tricky and extremely difficult to realize. The very hot plasma has a tendency to leak out of the bottle. The interaction between the plasma and the confining magnetic fields is not stable and scientists found many plasma instabilities: the building of large quantities of hot plasma was collapsing. Independent experiments in the USSR and in the USA led to the same disappointments. As the prominent thermonuclear pioneer and father of the US hydrogen bomb, Teller, pointed out about the Sherwood Project: 'It sher wood be nice if it worked.' The secrecy that had been imposed in the West and East alike was put aside at the second Atoms for Peace Conference in 1958. The magnetic confinement of plasma in magnetic bottles became the subject of an international collaboration. Today, this subject is an international effort with leading laboratories in the USA, Japan and the Soviet Union and a European joint project (JET) all working in collaboration. Considerable progress has been made in understanding the plasma confinement and in limiting the effects of the 'zoo' of plasma instabilities.

The United States declassified its laser fusion program in 1972 with the publication of a paper in the famous English scientific journal *Nature*. The Livermore scientist John Nuckolls and his colleagues wrote: 'Hydrogen may be compressed to more than ten thousand times liquid density by an implosion system energized by a high-energy laser. This scheme makes possible efficient thermonuclear burn of small pellets of heavy hydrogen isotopes and makes feasible fusion power reactors using practical lasers.'

Today, there is an international collaboration in 'small' thermonuclear devices triggered by huge lasers or gigantic particle beams. However, this collaboration is still limited because the basic physics of this approach is based on computer programs which were used for thermonuclear weapons. As the study of this subject becomes more developed and declassification more abundant, the inertial confinement approach to nuclear fusion is progressing.

The task of achieving controlled thermonuclear energy for solving the energy problems for all the generations ahead is enormous, difficult but not impossible. Plasma science and technology is the main issue of this international project and hope. During a 1987 meeting between the President of the United States, Ronald Reagan, and the Soviet leader, Michail Gorbachev, a decision was reached to collaborate in achieving controlled thermonuclear fusion energy during this generation. The Japanese government declared 'the achieving of fusion energy' as a national goal. These indeed are big steps in a positive direction, but much more effort and cooperation is still needed. But how did all this begin?

6.10 The Beginning of Controlled Nuclear Fusion in the USA

Following the Second World War, in the midst of pursuing the development of the hydrogen bomb (the uncontrolled nuclear fusion project), the American Atomic Energy Commission supported research on controlled nuclear fusion for energy application. To quote one early fusion pioneer, the American physicist Richard Post, 'once we have learned how to tap it, fusion can supply men's needs for energy for thousands of millennia, until, and even after, the Sun grows cold'.

On March 24, 1951, Argentine dictator Juan Peron called a news conference announcing that Argentina had a working nuclear fusion power plant. This scoop spurred some US scientists into the 'racing arena' for nuclear fusion research.

Already during the Second World War, Edward Teller and Enrico Fermi speculated about a controlled fusion reactor but most of the American scientists were not aware of this because of the secret classification of this subject.

Two prominent American physicists from Princeton, Lyman Spitzer and John Wheeler, were recruited by the US government to work on a secret project, called the 'Matterhorn Project'. Their assignment was to help Edward Teller and his team at Los Alamos, New Mexico, with the production of the hydrogen bomb.

Following Peron's dramatic announcement that Argentina possessed a working nuclear fusion reactor, Spitzer began to think seriously about controlled nuclear fusion. He asked himself how a hydrogen gas at high temperature (many millions of degrees, like the hot temperature in the center of the sun) can be confined without touching the walls. This question bothered him because no material can sustain such high temperatures. He reached the conclusion that the plasma can be confined by magnetic fields without touching the magnets or any other material. Once Sptizer thought that he knew how to confine the plasma, two new vital questions disturbed him: How could he create the plasma? How could he heat it? Soon he was able to solve these two problems by creating the plasma in an electric discharge and heating it by microwaves. But it appeared that as soon as he was able to solve one problem, another one crept up. For example, he could not measure the temperature of the plasma without disturbing it.

Over the next three years the American Atomic Energy Commission gave Spitzer one million dollars to pursue the controlled nuclear fusion project. A year after he began working on the Matterhorn Project, in the fall of 1952, the first Stellarator (Spitzer named the device Stellarator meaning a miniature star) was working. A Stellarator is a vacuum chamber in the shape of a doughnut surrounded by magnetic fields. Experimenting with Stellarator A, Spitzer estimated the temperature to be half a million degrees. In all his experiments, to his surprise, the plasma disappeared almost immediately, in less than a thousandth of a second.

Spitzer and his team felt that they needed a larger vessel with bigger magnets. Moreover, they soon realized that a cleaner gas is needed since the impurities inside the vacuum cooled the plasma. A second model, Stellarator B, began to operate in 1954. This time they achieved a temperature of one million degrees Celsius but again the plasma disappeared in a thousandth of a second, running to the walls of the Stellarator. They reached the conclusion that something was lacking in the theory. Edward Teller suggested that 'chronic' instabilities were responsible for the immediate disappearance of the plasma.

Determined to build a larger model and supported by Edward Teller, Spitzer received 20 million dollars to build a prototype fusion reactor, Stellarator C, with a temperature of 100 million degrees Celsius. In spite of the difficulties and uncertainties, other magnetic confinement devices were built in Livermore, Northern California, in Los Alamos, New Mexico, and in Oakridge, Tennessee.

Governing plasma for nuclear fusion energy and putting it to use is one of the greatest scientific and engineering challenges undertaken by man. Thus realizing that the American scientists alone could not solve the extremely difficult problems of controlled thermonuclear fusion, Teller and Spitzer asked the government to declassify this project. The American Atomic Energy Commission refused.

6.11 The Beginning of Nuclear Fusion in Britain and the Soviet Union

In the spring of 1956, the Russian Physicist Igor Kurchatov, the Head of the Soviet Union Nuclear Weapons Program, attended a conference at Harwell Atomic Energy Research Establishment, west of London. He was part of the Soviet delegation to Britain headed by the Communist party leader Nikita Khruschev and Soviet Premier Nikolai Bulganin. Kurchatov delivered a lecture entitled 'The possibilities of producing thermonuclear reactions in a gas discharge'. The British were stunned to learn that the Russians had been working on controlled nuclear fusion for peaceful means for the past six years. From Kurchatov's lecture it was learned that in the Soviet Union this research was initiated by the physicist Igor Tamm (who received the Nobel prize in physics in 1958) and a young 28 year old scientist Andrei Sakharov (who received the Nobel prize for peace many years later).

The Russian model called the Tokamak, built under the leadership of the famous Soviet scientist Lev Artsimovich, was different from Spitzer's Stellarator. In both the American and Russian scheme, the vessel containing the plasma was in a doughnut-shaped ring. In the Russian model, a plasma current was induced via a transformer inside the doughnut. This current creates its own magnetic field to confine the plasma. Here the Russians used the same magnetic confinement principle as used by the British six years previously. However, the Russian version was improved by adding an extra and crucial magnetic field in a perpendicular direction to that created by the circular current.

As already mentioned above, the first to begin research in controlled nuclear fusion were the British. As far back as 1946, Sir George Paget Thomson (the only son of J. J. Thomson) and Moses Blackman filed a secret patent application for a doughnut-shaped device that they had designed in Imperial College of Science in London. Sir George P. Thomson (who received the Nobel prize in physics in 1937) was also involved in the Manhattan Project for developing the US nuclear bomb in Los Alamos.

Now let's go back even farther. In 1919 in Cambridge, the physicist Francis William Aston invented the mass spectrometer, a device that can separate and measure isotopes. He was awarded the Nobel prize in chemistry for this invention in 1922. With his spectrometer, he was the first to measure isotopes. He found that the mass of helium is less than the mass of its components (two protons and two neutrons). This implies that by fusing lighter elements into heavier elements, the missing mass is transferred into energy (using the famous Einstein formula $E = mc^2$).

A year later, the famous British astronomer, Sir Arthur Stanley Eddington, gave a lecture suggesting the possibility that nuclear energy is locked in the atom and could be used.

In 1929, nine years after Eddington's suggestion, another British scientist, Robert Atkinson, together with the German scientist, F. G. Houtermans, wrote a paper that is considered the beginning of thermal nuclear energy research. They were the first to suggest that the energy of the Sun comes from thermonuclear fusion reactions.

In 1934, the famous New Zealand-born scientist, the father of nuclear physics, Ernest Rutherford (who received the Nobel prize in 1908), together with M. Mark Oliphant and P. Harteck, demonstrated in the laboratory the fusion of deuterium nuclei releasing a significant amount of energy. They used an accelerator to accelerate deuterium to hit a deuterium target. This was the first experiment to prove that nuclear fusion is possible. However, accelerators cannot be used efficiently for commercial power production because, when accelerating a particle, more energy needs to be invested than can be received in a fusion reaction.

In January 1947, following Thomson's and Blackman's application of a year earlier, John Cockroft (who received the Nobel prize for physics in 1951), the head of the British Atomic Energy Research Institute at Harwell, held a meeting to discuss a controlled fusion program where Thomson explained his ideas. Present at the same meeting was the Head of the Theoretical Physics Department at Harwell, Klaus Fuchs, who was one of the British team who collaborated with the Americans on the atomic bomb in Los Alamos. Years later he was convicted as being a Russian spy and received a jail sentence. Also present at the meeting was another Harwell member, an Italian scientist Bruno Pontecorvo, who later was also found to be a Russian spy. He managed to escape to the Soviet Union before he was caught.

Did Fuchs and Pontecorvo pass the important information at the meeting of 1947 to the Russians and is this how the Russians began working on the Tokamak? It is conceivable to assume that the early research on controlled and uncontrolled fusion in Britain was reported to the Soviet Union.

Following the above meeting, Thomson's and Blackman's proposal was not supported financially by the British authorities. However, controlled nuclear research was not neglected in Britain.

6.12 International Declassification of Controlled Nuclear Fusion

The scientists who were working in the 1950s on controlled fusion research in Great Britain, the Soviet Union and the United States were the same ones who during the Second World War were building nuclear weapons.

Similar to their colleagues in the US, the British and Soviet scientists were running into many difficulties in building the magnetic devices for controlling nuclear fusion and were in favor of declassifying this subject.

Following the Soviet successful launching of the first satellite into orbit in 1957, the *Sputnik*, the Americans agreed to declassify the subject of controlled thermonuclear fusion. In 1958 in Geneva, during the United Nations Conference on the peaceful uses of atomic energy, the magnetic fusion research in the United States, Great Britain and Soviet Union was declassified.

On the eve of declassification, John Cockroft, the director of Harwell Atomic Research Institute in England, gave a press conference saying that a circular pinch machine, ZETA (Zero Energy Thermonuclear Assembly) produced plasma temperatures of 5 million degrees with a plasma confinement of 3 thousandths of a second. Thus it was proclaimed that Britain was the first in the world to achieve a controlled fusion reaction.

Both the well-known Soviet experimentalist, Lev Artsimovich, and Lyman Spitzer, the leader of the US program, disproved the British findings and claimed that this was not fusion but neutrons that were detected as a result of the acceleration of deuterium particles on to other deuterium particles. Therefore the plasma temperature was not achieved as claimed by the British.

A few months later, Cockroft issued a press release saying that the reactions produced from the ZETA were not caused by true fusion reactions, confirming the above criticism.

The Geneva Conference in 1958 marked the birth of the world fusion community and the beginning of international collaboration. Despite the cold war, the exchange of scientists between East and West emerged.

The British–Russian collaboration began between Lev Artisimovich and Sebastian Pease, the director of the British Cullham Laboratory near Oxford. Having convinced their governments, in 1961 a contract for collaboration between Britain and the Soviet Union was signed. The same year the first International Atomic Energy Agency Commission Conference in Salzburg, Austria, on controlled nuclear fusion was inaugurated.

In 1966, Lyman Spitzer, the founder of the American fusion research, retired from the thermonuclear plasma research and returned to work in astrophysics.

Two years later, in 1968, Artsimovich announced that the plasma inside his Tokamak reached a temperature of 10 million degrees during 100th of a second. Remembering the British fiasco ten years previously, his news was taken skeptically. In order to prove his point, Artsimovich invited a British team to come and measure the temperature to verify his claim.

The British delegation to Moscow was headed by Nick Peacock. Equipped with a computer, five tons of equipment and a laser, the stage for the crucial experiment was set. The British used a laser for the first time in order to measure the temperature of the plasma without disturbing it. This novel and accurate experiment proved that the temperature claimed by Artsimovich was indeed correct.

Following the positive Soviet results, scientists all over the globe began building Tokamaks as the main magnetic confinement devices. The entire American fusion program was revised and even at Princeton, the Stellarator was dismantled and a Tokamak was built. Soon bigger Tokamaks were constructed. The two largest devices, one in Princeton, USA, the TFTR (Tokamak Fusion Test Reactor) and the second the JET (Joint European Tokamak) in Cullham, UK, burned deuterium–tritium reaching 'break-even' nuclear fusion. These results indicated that the *output* nuclear fusion energy was about equal to the *input* electrical energy invested to heat the plasma. Although this 'break-even news' in the late 1990s was toasted with enthusiasm by the scientific community this result is still far from achieving an economically viable *high energy gain* nuclear fusion energy reactor.

So what is the next step? An international collaboration began a few years ago by the United States, Russia, Japan and Europe. Upon completion of the planning and design of the ITER (International Thermonuclear Experimental Reactor) in 1998 the United States decided to withdraw its collaboration. Some doubts were raised by the Americans as to the right pathway for solving the controlled nuclear fusion problems. The Americans believe that other nuclear fusion research concepts should be considered before larger reactors are built.

Today, the Princeton TFTR is shut down and the Tokamak research has been reduced drastically in the USA. With a more optimistic and stable policy, Europe, Russia and Japan are continuing their ongoing research on this subject. Furthermore, they are also considering other magnetic fusion concepts. Some scientists think that Sptizer's Stellarator was not such a bad concept and different versions of this concept are being again considered.

6.13 Landmarks in the Development of Plasma Physics

18th and 19th centuries		Development of electromagnetism.
1621	P. Gassenow	Named the northern aurora 'aurora borealis'
1750s	Benjamin Franklin	Scientific observation of lightning.
1773	Captain J. Cook	Named the southern aurora 'aurora australis'

1808	H. Davy	DC arc discharges.
1830	M. Faraday	Electrical discharges in tubes; introduced (from Greek) the terms electrode, anode, cathode, ion.
1860	J. Plucker	Detected the fluorescence phenomenon induced by cathode rays. Diverted the trajectories of cathode rays with magnetic fields.
1864	J. C. Maxwell	Introduced the complete equations of electromagnetism.
1874	G. Stoney	Suggested the existence of the electron.
1879	Sir W. Crookes	Research in low density discharge tubes. Coined the Fourth State of Matter.
1882	B. Stewart	Suggested the existence of an electrical conducting layer in the atmosphere.
1883	T. Edison	Induced a current in a vacuum tube by heating one of the electrodes.
1896	A. H. Becquerel	Discovered radioactivity.
1897	J. J. Thomson	Discovered the electron.
1901	G. Marconi	First electromagnetic wave transmission across the Atlantic Ocean.
1905	Albert Einstein	Special theory of relativity and suggested that mass and energy are equivalent.
1906	H. A. Lorentz	Coined the term electron and explained the electric and magnetic forces acting on electrons.
1906	Lord Rayleigh	Collective oscillatory behavior of electrons.
1908	K. Birkeland	Relation between laboratory plasma and cosmic plasma.
1909	G. Claude	Invented the neon tube.
1911	R. A. Millikan	Measured the charge of the electron.
1911	W. Gaede	Developed the mercury diffusion pump after his invention of the rotary mercury pump in 1905. This enabled lower densities to be reached in a gas tube.
1913	Niels Bohr	Developed the theory of the atom. The father of atomic physics.
1920	E. Rutherford	Experimental proof of Bohr's theory. The father of nuclear physics.
1923	P. Debye	Developed the theory of dissociation of electrolytes.

1926	M. A. Tuve and G. Breit	Experimental proof of the existence of the ionosphere.
1928	I. Langmuir	Introduction of the term plasma for the Fourth State of Matter.
1929	I. Langmuir and L. Tonks	The beginning of modern experimental plasma physics.
1929	R. Atkinson and F. Houtermans	The beginning of research on thermonuclear fusion. Explained the solar source of energy by fusion of hydrogens.
1933	G. Claude	Constructed the first practical fluorescent lamp.
1934	W. H. Bennett	Suggested the plasma pinch phenomenon.
1934	M. H. Oliphant, P. Hartech and E. Rutherford	First experiment of nuclear fusion in a laboratory (deuterium–deuterium).
1936	L. D. Landau	Introduced the statistical theory of plasma.
1938	H. A. Bethe	Determined the thermonuclear cycle of the Sun.
1938	O. Hahn, F. Strassmann and L. Meitner	Achieved nuclear fission of uranium.
1942	H. Alfven	The beginning of hydrodynamics to describe plasma physics. Introduced the equations describing the motion of plasma as a fluid in electromagnetic fields.
1942	E. Fermi *et al.*	First nuclear fission reactor.
1945	J. R. Oppenheimer *et al.*	First atomic bomb explosion.
1946	L. D. Landau	Described the interaction between particles and plasma waves. The beginning of modern theoretical plasma physics.
1947	Sir E. V. Appleton	Discovered the upper region of the ionosphere.
1950s	C. H. Townes, N. G. Basov and A. M. Prokhorov	Suggested the theory of a laser (awarded the Nobel prize in physics for this work in 1964).
1951–52	E. Teller *et al.* and A. Sakharov *et al.*	The explosion of the thermonuclear fusion devices known as the hydrogen bomb.
1952	L. Spitzer	Built the first magnetic confinement device, the Stellarator.
1957	J. D. Lawson	Introduced an energy balance criterion for gain in a nuclear fusion reactor.

1958	J. A. Van Allen	Discovered the radiation belts by *Explorer I* satellite, later referred to as Van Allen Belts.
1958	Under auspices of the United Nations	The Second International Conference on the Peaceful Uses of Atomic Energy; declassification of the controlled thermonuclear projects. The beginning of the modern era in plasma physics.
1960	T. H. Maiman	Constructed the first laser system.
1962	NASA (National Aeronautics and Space Administration)	Established the existence of the solar wind.
1968	N. Basov *et al.*	Demonstrated fusion reactions by focusing a neodymium glass laser beam on a lithium–deuterium solid target.
1968	L. Artsimovich	Developed the Tokamak (originally suggested by I. Tamm and A. Sakharov).
1972	J. Nuckolls *et al.*	Suggested the idea of high compression in inertial confinement fusion (this was the first USA declassification of the subject).
1970s		The development of the plasma electronic microelectronics industry.
1990s		Processing of industrial wastes with plasma.
1993	USA	Declassification of the indirect drive scheme in inertial confinement fusion.
1996–97		The US Princeton Tokamak (TFTR) and the European Tokamak (JET) burned deuterium–tritium fuel to achieve 'break-even'.
1997	Sandia National Laboratory, USA	Two million joules of X-rays in a few nanoseconds were achieved in a Z-pinch device.

Appendix

Rhyming Verses

These rhyming verses were an integral part of the first edition of *The Fourth States of Matter*. Many readers congratulated us for the 'refreshing' and 'original' verses. The rhyming verses are by no means intended as poetry, nor do they follow any specific parameters, patterns or metrical forms. The purpose of these rhyming verses is to put big ideas and complicated issues into a compact, simplified and sometimes easy to remember form.

Matter 1

Matter is something very real,
Matter is everything man can survey, touch or feel.
The stars, the seas, the trees, the ground,
Matter is everything all around.

Matter is the air we breathe, it's life itself,
Matter is the water we drink, it's man himself.
Matter is the homes we dwell in, the food we eat,
Matter is up in the atmosphere, or under our feet.

Matter is whatever occupies space,
Matter is substance in each different phase.
Be it solid, gas or water,
Everything in the Universe is matter.

Matter 2

Democritus believed that there exists a matter so unchangeable,
That the atoms of each particular substance are indestructible.
Some of these substances are made of fundamental particles of matter
And others are the result of the union with the latter.

But his ideas of elements and compounds Aristotle thought absurd,
He believed that there were only the elements, fire, water, air and earth.
It was the combination of some substance with one of his four,
Which determined the nature of any particular substance and not more.

Following Aristotle's teachings were thousands of alchemists, or even more,
Mixing and brewing and hoping to get gold out of his four.
Though today alchemy is referred to as a mere fool's search for gold,
Their persistent testing and examinations did much unfold.

Finally Democritus' vision of the atom was resurrected,
First by Galileo, then by Dalton it was corrected.

Matter 3

The Greek philosophers who questioned why
For their persistence did sometimes die.
But shrewd and fearless they did persist
What is the Universe, why does man exist?

Some claimed it was water because it was visible
Others said air, earth, or fire or something invisible.
After 2000 years we refer to the latter
And know today the three states of matter.

The solid such as the rock found in the mountain,
The liquid as the water flowing out of the fountain,
The gas the invisible oxygen in the air—
These states of matter can be found everywhere.

The fourth state of matter, the plasma, is here on Earth so scarce,
Yet infinite quantities can be found in outer space.
The Sun, the stars, almost all the Universe about
Are in the form of plasma, no doubt.

Matter 4

Nuclear physics is the science of the protons and neutrons,
Atomic physics is the science of the electrons.
The electrons are moving around the nucleus one by one,
In a circular motion like the planets move around the Sun.

The electrons are racing around the nucleus due to the electric forces
And are determined in their attraction to remain on their courses.
In order to pull an electron out of a trajectory,
Once can coax him out by investment of energy.

This energy can be invested by a photon,
Or by a collision with another electron.
The photons are part of the electromagnetic radiation,
And play an important role in the plasma creation.

The Four Forces

The four types of force provided by nature,
Provide the laws of the Universe, its legislature.
Through their different fields they rule our existence,
Either attracting or repulsing, though with persistence.

The gravitation field with its pulling action
Is an all-pervading one which exerts an attraction.
An attraction drawing objects to the center of the Earth,
And causes the weight that the objects exert.

The electromagnetic field, the source for all chemical binding,
Causes electrons to atoms and atoms to each other in uniting.
Besides an attraction, this force possesses the repulsion trait,
And is an important phenomenon in the research for the plasma state.

The strong nuclear force is short ranged by far,
But inside the nucleus it's the leading star.
It holds the nucleus together and is therefore its hero,
Outside the nucleus, the force falls rapidly to zero.

The weak interaction is responsible for the natural decays
Of all radioactive matter that emits beta rays.
This force with electromagnetism was unified, therefore,
We now refer to three types of force only and not four.

To unify gravity with electromagnetism was Einstein's hope,
Though try as he did, with this theory he couldn't cope.
Today scientists' dreams and determination
Are to make all the four forces a unification.

Electricity

Electricity is an invisible force of nature provided to man
To help him produce light, heat and energy, as best as he can.
Electricity can produce charge in different kinds of matter
Whether it's in the form of solid, gas or water.
Electricity exists as positive and negative states,
Producing electric currents to flow at different rates.

Flowing across a wire inside a light bulb or another fixture,
The current is composed entirely of an electron texture.
Flowing inside an electrolyte solution variation,
The current is composed of positive and negative ion delegation.
Flowing inside a vacuum tube certain rays are detected,
Composed of electrons and positive ions that are ejected.

The current flows through a closed circuit of metal wire or other matter
Beginning at one pole and ending at the opposite of the latter.
This electric current which carries an electric charge so fine
Is made up of streams of charged particles flowing in a line.
As the electrons are easy to pull out of their pass,
It was concluded that the electrons have little mass.

Radioactive Materials

Radioactive materials give off three kinds of radiation,
Alpha, beta and gamma rays, are their identification.
The first two always electric charges haul,
While the third carries none at all.

The positive electric charge is carried by the alpha,
The negative electric charge is carried by the beta.
Both are streams of charged particles, smaller by far,
Even smaller than the tiny atoms are.

The alpha's positive electric charge
Is twice as great as the electron's negative charge.
The alpha particle, the nucleus of the atom of helium gas,
Has about four times the mass the proton has.

The third, the gamma, which are continuous rays,
Are lightlike radiation made up of waves.
Waves much shorter than those of ordinary light,
That penetrate through or bounce to the left or right.

Following many experiments and revisions,
Scientists reached the following decisions:
The beta particles, identical to the electron in mass and charge,
Are actually like the electrons spinning around the atom at large.

Thus scientists had assumed the atom to be such a bit
So extraordinary small, that it cannot be split.
Yet the alpha and beta particles that came to light
Showed that the atoms could be made to disunite.

The Isotope

The isotope is actually merely a name
Of two or more atoms, which are the same.
Possessing equal number of protons and electrons,
Yet varying in the number of the neutrons.

Radioisotopes possess the same regulation,
Except for the fact that they emit radiation.
Many uses today and requirements they meet,
Whether in medicine, industry or as a source of heat.

The Atoms

Atoms of varying elements their identity sustain,
By the number of three kinds of particle they contain.
In the nucleus of the atom dwell the protons and neutrons,
Spinning around at breakneck speed race the electrons.

The atom is so small that one sand grain
Can millions of billions of atoms contain.
The core, though one hundred thousand times less in size,
Yet most of the atom's mass does comprise.

In each atom, the number of electrons
Is the same as the number of protons.
The number of neutrons in each atom can vary
Except the hydrogen atom, not one does it carry.

The proton carries the charge which is the positive,
Whereas the electron has a charge which is negative.
The electron also has a very light mass
About 2000 times less than the proton has.

The third, the neutron, with no electric charge whatever,
Has a mass about the same as the proton, however.
The discovery of these three different elements,
Led science on forward towards newer developments.

Ion

Ion, from the Greek verb meaning 'to go',
Causes electricity through solutions and liquids to flow.
Ions are atoms that have acquired an electric charge,
From the electrons that are spinning around at large.

If one or more electrons from the atom are pulled out,
The result is a positive ion, no doubt.
If one or more electrons to the atom are increased,
The result is a negative ion released.

From First to Fourth

In the beginning, there was nothing at all,
Except for the Universe, compressed in a ball.
Then followed the 'Big Bang' episode,
Causing the ball to violently explode.
Creating entirely a hot plasma lather,
Which was actually the first state of matter.

After the explosion, the matter cooled down,
Turning the fragments into stars, just like our Sun.
During the expansion of this gigantic heat storm,
Matter cooled down to the gas form.
The cooling continued to liquid and solid formation,
To form the different matter configuration.

In the beginning, with the coming of civilization,
Prehistoric man's first inauguration
Was to the earth and the rocks; the rain and the water.
These solids and liquids became the first and second states of matter.
The gas was realized later and became the third phase,
And last came the plasma to interchange from first to fourth place.

Plasma

The meaning of plasma in medicine is quite ordinary,
A simple explanation can be found in any dictionary.
It's the watery fluid of organic compounds and mixtures,
In which the cells of the blood are suspended as permanent fixtures.

Plasma in science is not at all the same,
Although mistakenly it acquired the exact name.
In blood plasma is a liquid, similar to water,
In science plasma is a gas and the fourth state of matter.

A cube of ice which is cold matter,
When heated up turns into water.
This liquid or water when heated some more,
Results in the production of a steam or vapor.

This vapor or gas when heated to extreme degrees
Results in a new type of gas release.
This new gas is the plasma creation
Whose process in physics is called IONIZATION.

The inhabitants in the solid, liquid and gas phase
Are moving molecules or atoms from place to place.
In the plasma state configuration,
The electrons and ions are in firm domination.

The electric forces govern the mobility
Of the different particles' CONDUCTIVITY.
While the gas is not a good candidate,
The plasma is a conductor of the 'top rate'.

The Coulomb force which is long ranged,
In the plasma scene becomes changed.
The range is shortened to the DEBYE LENGTH,
Since the shielding has weakened the force's strength.

The COLLISIONS in a plasma between its population
Are due to the Coulomb force dictation.
These probabilities are described by CROSS SECTIONS,
Which are the effective areas for the connections.

Plasma is a quasineutral combination
Of mobile charged particle correlation.
The collective behavior in plasma is related,
To the MANY WAVES that are created.

The gases of ion and electron compositions
Are achieved under very hot conditions.
Creating and containing the plasma depends on many aspects,
Temperature, density, INSTABILITIES and other effects.

Ignorance of plasma in science is due to the conclusion
That most of the educators prefer its exclusion.
Yet as most of our Universe is in a plasma form
Plasma in science should become more known.

The study of plasma is under serious investigations:
To develop cleaner and more efficient INDUSTRY APPLICATIONS,
To understand the UNIVERSE OF PLASMA situations
And to obtain FUSION ENERGY for the future generations.

The Big Bang

The Big Bang model became more accepted,
As new experimental facts were detected.
Scientists are pleased with the new revelations,
As they fit in well with their explanations.

When the astronomers observed the outer space,
They found some photons out of place.
These photons were a uniform background radiation,
That didn't seem to arrive from any constellation.

That background radiation had a temperature so low
Of three degrees above zero and lacked any glow.
Thus it was concluded that this left-over content
Was the remains of the Big Bang event.

Hubble had long concluded that the galaxies tend to recede,
Drifting apart at a very high speed.
The expanding of the Universe in the Big Bang interpretation
Fits in well with Hubble's previous explanation.

The strange composition of a star
Is the same as predicted by Big Bang so far.
The stars and galaxies have a content
Three times more hydrogen than the helium element.

The Big Bang Model is consistent with geologists' insistence
That the Earth is about ten billion years in existence.
Thus scientists today are pleased with the Big Bang suggestions
As for the time being they satisfy any of the above questions.

All the Stars

All the stars that we observe in the sky
Are first born, then develop and finally die.
Their birth is formed from the gas, the plasma and the grains,
Stemming from the residues of the Big Bang remains.
Forming a massive cloud with strong gravitational domination,
Causing the matter to condense and a temperature escalation,
Thus providing the ideal situation
For thermonuclear reaction continuation.

The continuous burning of the nuclear fuel for a length of time
Results in the radiation of the stars and their bright shine.
The nuclear burning is the time of their development
When the hydrogen is changed to the helium element.
The death of the star can usually depend
On its mass and whether it is alone or has a friend.
The star can collapse and take up a new role,
Either as a white dwarf, a neutron star or as a black hole.

A Supernova

In 1987, on the night of the 23rd of February,
Scientists were privileged to witness something extraordinary.
A supernova had come to play out its final role,
As a neutron star or as a black hole.

A white dwarf follows the trend
Of collecting mass from his friend.
Reaching a mass forty per cent more than our Sun,
It explodes as a supernova of type one.

Large stars containing extremely big mass
Eventually become a supernova of the type two class.
But before reaching this last stage
They fuse from hydrogen to the 'iron age'.

The Aurora

Due to Sun storms or some other disturbance,
A burst of charged particles descends from the turbulence.
Colliding with atmosphere atoms and causing separation,
Thus releasing electrons and ions to form ionization.
The different shapes of the light creation
Are thus the result of the plasma formation.
This *aurora* in the northern hemisphere is the Borealis,
The one in the southern hemisphere is the Australis.

The Ionosphere

The first radio signals received across the Atlantic Ocean
Created a great uproar and world-wide commotion.
The atmospheric layer of the plasma composition
Enabled the radio signals' transmission.
This layer formation, known as the ionosphere,
Soon explained the radio transmission quite clear.

The streams of photons arriving from the Sun
Were absorbed by the atoms one by one.
With energy to remove the electrons from the atom's configuration
Enabled the atom to produce the plasma layer formation.
This plasma thus reflecting the waves that swerve
Allowed the radio waves to follow the Earth's curve.

A Very Short History of Plasma

With Faraday's electrical charges that came to pass,
A new glow appeared, different from those in solids, liquids or gas.
Sir William Crookes while observing his own studies,
Of the electrical charges found in these gas bodies,
Suggested this glow so different from the other
Should be known as the fourth state of matter.

This new state of matter could not be understood
Before the following developments and events occurred.
Thomson's discovery of the electron and its nature,
Bohr's and Rutherford's completion of the atom's structure.
Langmuir's and Tonks's continuation of Crookes's experimentation
Of the gas discharge properties and the new creation.

The name plasma was Langmuir's own initiation
To describe the electron and ion correlation.
Faraday's work on solutions and those unanswered queries
Were summarized and brought to light by Debye's new theories
Of ions in solutions, a theory so profound and outstanding
That enabled the continuance of the plasma understanding.

Plasma motions as a fluid in a magnetic field was Alfven's addition
And he was awarded the first Nobel Prize in plasma for this contribution.
The famous Russian physicist Landau and his sophisticated construction
Of his theory of the electron and the plasma interaction
Paved the way for the modern plasma observations,
Through his valuable achievements and revelations.

Science ('Never the less')

Science, in the beginning, was a mere curiosity,
Based on hypotheses and sophisticated philosophy,
With no experimentation theories were hard to contradict,
And hindered new concepts that others dared to predict.

Science, in the beginning, was to think and to ponder,
To search the skies and surmise what lay yonder.
Those ancient astronomers who lacked the simplest of lens,
Based their theories on what the human eye could see and on their
    common sense.

Ptolemy's doctrine that the Earth was the center, fixed and unmovable,
Was adapted by the Roman Catholic Church and was thus indisputable.
That the Sun and planets circled around the Earth was most beneficial,
As it made the Earth supreme and the church's leader its official.

Copernicus's theory that the Sun was the center and not the Earth
Was banned by the Church and thought to be absurd.
That the Earth and planets circled around the Sun was not taken seriously,
And that the Earth turned on its own axis was attacked furiously.

Bruno courageously the Copernican theory did brace,
Adding that other inhabited planets were whirling about in space.
His teachings and beliefs he refused to forsake,
He was tried by the Inquisition and burnt at the stake.

Brahe and Kepler continued the observations
Of the heavenly bodies and their interpretations.
Kepler's greatest achievement was his genius conclusion
For the movement of the heavenly bodies in a simple mathematical
solution.

Perhaps the greatest astronomer of those early days
Was Galileo, whose contributions enlightened science in many ways.
With the magic of the telescope he made new revelations
Of the surface of the Moon and the different constellations.

He followed the Copernican theory that he now could prove,
But was forced to retract that the Earth does move.
With his creative ability, however, for experimentation,
He opened up a new era of discovery and exploration.

Sir Isaac Newton (1642–1727)

In 1642, the same year that Galileo met his death,
In Woolsthorpe, Isaac Newton took his very first breath.
As a youth he was not an especially good student,
But he taught himself mathematics and was most prudent.

From Kepler's book on optics, he learned about physics and astronomy,
From an ordinary chemist, he learned to experiment in chemistry.
Before he was 24 and while still a student in Trinity,
He laid the basis for his great discoveries and natural ingenuity.

His principles of calculus stirred a great commotion,
It was a new system of mathematics for calculating motion.
Experimenting with a prism with which he spent many an hour
He worked out his advanced theory of light and color.

Returning to Woolsthorpe in 1666 to escape the horrid plague all around
It was there legend says that he observed the apple falling to the ground.
Whether the story is true or not, however, thereafter very soon,
He began to think of gravity extending to the orb of the Moon.

Being a cautious man he hesitated to publish what he did discover,
And continued with his theories involving light and color.
Finally, his ideas on forces and motions were publicized
In the book the *Principia*, which soon became well known worldwide.

His publishings brought him esteem and enthusiasm
And many foreign academies competed to honor him.
He was appointed Master of the Mind and President of the Royal Society
And in 1705 was knighted by Queen Anne for his ingenuity and loyalty.

Newton had a tremendous influence on theoretical and applied science,
And some of his discoveries had a direct appliance.
With calculus the problems of curves were reduced,
Which led to improved weapons which were then produced.

The tides could be explained by his theory of gravitation,
Which helped in map making and aided navigation.
The improved weapons and navigation led to new explorations,
And strengthened Western Europe over other world nations.

The ingenuity of calculus paved the way to new revelations
Of handling curves, variations and differential equations.
His work on mechanics was supreme and without flaw
Even today we refer to the physical universe as governed by Newton's law.

Throughout his life he shunned publicity,
And retained a modest life of simplicity.
He died in 1727 at a ripe old age of 85,
But his numerous contributions will forever survive.

Energy

Energy is responsible for our 'easy life',
Without energy civilization cannot survive.
Energy provides a multitude of things such as light and heat,
It enables man to grow his food so that he could eat.

Energy is the capability of doing toil,
Today most of our energy supplies come from gas, coal and oil.
As these raw materials are not in perpetuity,
Nuclear fusion must supply the energy continuity.

Fission and Fusion

Nuclear fission can create an explosion
That can bring about chaos and erosion.
Using uranium, so hazardous, for its construction,
The atomic bomb was created for destruction.

Nuclear fusion can be made to explode
Together with the atomic bomb caused much more to erode.
Using deuterium found in oceans all about,
The hydrogen bomb is much more harmful, no doubt.

For today's civilization and peaceful means, however,
Fission and fusion are meant for something better.
To solve the problem of the energy situation,
For the present and future generations.

Thus nuclear fission reactors were constructed on condition
That the energy release would be under controlled supervision.
But working with uranium is 'dirty', or so the media has made
widespread;
As it emits radiation, it's become a psychological dread.

And so nuclear fusion, the scientists today believe,
Could be the answer to mankind's relief.
Using deuterium so 'clean' and found with ease,
Can provide an abundance of energy release.

A Magnetic Bottle

A magnetic bottle is so different than any other,
That contains milk, cola, or ordinary water.
Its shape and construction can be so bizarre,
That can resemble a doughnut or a monstrous car.

As the plasma is created inside,
And must be prevented from leaking outside,
Those magnetic bottles were so designed
In order that the plasma could be confined.

The plasma inside the bottle must be very hot,
But how can these mischievous particles be caught?
This is accomplished with large magnetic fields
Created inside the plasma which act as shields.
These magnetic fields can have a strange configuration,
And control the ion and electron gyration.

In nuclear fusion two light nuclei can be fused,
During the collision a heavier nucleus is produced.
For such a production large temperatures must be sustained,
So that nuclear reactions can be maintained.
The burning is accomplished by the constant repeating
Of either ohmic, neutral beam or microwave heating.

After plasma ignition is acquired,
The above heatings are no longer required.
However to ignite plasma is far from an easy assignment,
As the instabilities are determined to ruin the confinement.
Moreover, added to the problems is the diffusion trait
Which tries to dissolve the plasma state.

Although the scientist is faced with a difficult mission,
To achieve controlled fusion is his ambition.
He will continue with toil and dedication,
To find an energy source for generation after generation.

The Laser

What is a laser as a simple definition?
Light amplification of stimulated radiation emission.
Used in medicine, industry and research world-wide,
To improve the conditions of all mankind.

In medicine lasers are used in eye and other operations,
In industry as cutting tools and other applications.
In research to study the different phenomena of light variation,
Enlightening the atomic structure from the irradiation.

If there are sufficient electrons in the upper energy position,
The stimulating photons can further stimulate emission.
These photons cause the electrons to jump to a lower state,
Creating a coherent beam with a very definite energy rate.

Excessive excited electrons known as the population inversion
Are maintained by the power fed into the system for this transition.
Thus a chain reaction similar to that of nuclear fission
Can be achieved for the light stimulated radiation emission.

Today the laser can be worked in gaseous, solid and liquid states
Causing photons to stimulate electrons at high frequency rates.
For the future, however, higher frequencies are under investigation.
To include as well the biological structure exploration.

Inertial Fusion Energy

Powerful lasers or intense particle beams
Can compress a small pellet to such extremes
That the temperature inside the core will greatly increase
So that nuclear fusion processes cause an energy release.

A plasma is created around the pellet by lasers or beams
From the absorption of the photon or the ion streams.
Both the temperature and the pressure of the plasma are magnified
Inducing shock waves so that the compression is intensified.

If the instabilities can be constrained
A symmetrical compression can be obtained.
Thus when this pellet is greatly squeezed
Nuclear fusion energy is released.

The above process is terminated very fast
Since the fusion reactions do not last.
In order for this scheme not to be in vain
Energy initially invested should show a high gain.

Albert Einstein (1879–1955)

A man, a genius, an intellect, a philosopher and a physicist,
Celebrated as one of the world's greatest scientists.
His logic and clarity and creative imagination
Proved to all around him a great inspiration.

He was forced to attend a secondary school before he could enter,
To study at the famous Swiss Polytechnique Center.
He earned money through teaching and doing computations
And became the Swiss patent examiner upon a friend's recommendations.

It was during this time that Einstein was most active,
His numerous hours of research proved most productive.
In 1905 he produced three papers that revolutionized man's imagination
Of the physical universe, laying down the basis for the atomic age
foundation.

His emission of electrons from metal surfaces exposed to light
explanation
Theoretically was one aspect of the quantum theory inauguration.
In another he analyzed the Brownian movement theory mathematically
Providing a method for determining the dimensions of molecules
accurately.

The third, describing the relativistic nature of uniform motion,
And the interdependence of space and time, then caused a commotion.
The equivalence of mass and energy, his famous resolution,
$E = mc^2$, was dramatically demonstrated in the atomic explosion.

In 1915 he published his General Theory of Relativity,
In which he developed the geometry concept of gravity.
Einstein here achieved the unification
Of space, time, mass, energy, inertia and gravitation.

Einstein loved music and played his violin with zest,
He loved Beethoven, Bach but Mozart the best.
He believed in a world of harmony and simplicity,
He was full of benevolence, humor and integrity.

Though a confirmed pacifist and peace was his object,
He was responsible for initiating the atomic bomb project.
Fearing the Nazis would achieve world domination,
He urged President Roosevelt to pursue nuclear research investigation.

Albert Einstein is so well known,
We learn his name before we are fully grown.
As a scientist considered to be one of the best
His legacy left to man is a very rich bequest.

Epilogue

Although the writing of this book was stimulating and rewarding, it was far from an easy task. The collaboration between a person who speaks 'physics' and one who doesn't is, quite frankly, very exhausting. The gap between the physicist and the non-physicist is huge. To explain physics to a person lacking any background in mathematics is extremely difficult because the 'language' of physics is mathematics.

The fundamental laws of nature and their implications are written and understood through mathematics. Mathematics is not just a language; it is a tool for reasoning. Mathematics connects one statement to another; one law to another. The logic of mathematics links one fact to another. Without mathematics, the sciences are a huge collection of facts lacking any common basis. Mathematics enables us to develop consequences and analyze physical laws. However, in order to explain physics in this book to the layman, we had to pay the price of avoiding mathematics. By doing this, I feel that much of the beauty, the completeness and the excitement of physics was lost.

Mathematics is the necessary tool for understanding physics. Nonetheless, in physics a connection between mathematics and the real world must be established. After the calculations have been worked out theoretically, they must be incorporated into fact. Thus the final 'judge' of a physical law or its conclusion is the 'experiment'. Physics is not just abstract philosophy or 'intelligent argument' but a philosophy and logic which is related to the behavior of nature and confirmed by experiment.

How did I let myself be talked into undertaking a 'mission' of collaborating on a book in physics without using even the simplest mathematics? My enthusiasm was first aroused when my wife began to pump me with her endless questions. I tried patiently to explain the difficult phenomena, especially those concerning plasma physics, involved in the research being carried out at our department, where she had worked as a secretary. During some of my arguments and discussions with my fellow colleagues, she would eavesdrop and ask a most intelligent question. Many of my colleagues would often say that Yaffa should have studied physics.

Yaffa, who enjoys writing, compiled some rhyming verses in physics which I found delightful and simplifying. After she had written quite a few verses, she approached me with the following proposition: 'Let's write a simple book in English without using any of the complicated graphs, "big" and "small" numbers and equations.' At first I rejected her idea, insisting that there are certain phenomena in physics that cannot be explained without using equations or other 'physics aids'. But here I find myself in collaboration with my wife, and to quote her, 'we did it'.

We have tried to write a simple and clear book in plain English. I have not been able to find a popular and simple book on plasma physics with the constraints imposed by my wife. I have not even been able to find a reliable historical background on the development of plasma physics. I have compiled the plasma historical background from bits and pieces of different books and from bits and pieces that have been exchanged in scientific circles.

We have tried to include some picturesque situations to help the reader understand some of the difficult physics and established facts related to plasma physics. I am sure that we have strayed somewhat from a precise explanation with our analogies but I hope that scientists will not find these too 'imperfect'. The scientists do not need these analogies; they were introduced for the intelligent person who lacks a background in physics.

In our book we have tried to translate plasma science from the language of physics to the simple language of English. It is hoped that, in this way, this subject will be more comprehensible to the layman, the curious and intelligent individual, the student, the physicist's wife and any other curious reader. If people who do not speak physics understand this book or even some parts of it, we have accomplished our difficult mission. If, after having read this book, the reader becomes acquainted with plasma physics and realizes the importance of this subject for solving the energy problem, we think that a valuable message has been delivered.

I feel it appropriate to conclude my epilogue with the theme from the well known book *The Two Cultures and the Scientific Revolution* by the famous author and physicist, C. P. Snow. According to this book, there are two disciplines, 'Science' and 'Literature'; one discipline knows very little about the other. Moreover, communication between these two 'cultures' is very difficult and may be impossible. Our book is an attempt to disprove Snow's philosophy. Our attempt at 'invading the fourth state of matter' is a result of the communication between the two cultures, thereby making the scientific culture available to the public at large.

Shalom Eliezer
Rehovot, Israel
March 1988

Glossary

Ablation surface: Place where the laser irradiated target material is evaporated and plasma is created.

Absolute zero: The state where all atoms are at rest and therefore the lowest temperature theoretically possible. This temperature is the zero point of the Kelvin temperature scale and is equal to −273.15 degrees Celsius or −459.67 degrees Fahrenheit.

Acceleration: The rate of change of velocity with time in either magnitude or direction.

Accelerators: Devices which increase the speed and the energy of electrons, protons, or other electrically charged particles.

Alfven wave: A transverse wave (analogous to a vibrating string) in a plasma within a magnetic field.

Alpha particle or alpha ray: One of the particles emitted in radioactive decay. This particle is the nucleus of the helium atom which consists of two protons and two neutrons.

Ampere: The practical unit of measuring an electric current.

Amplitude (of wave): The maximum value of the displacement of a wave from its undisturbed position. In an electromagnetic field, this is the maximum value of the electric or magnetic field.

Anode: The electrode of a vacuum tube (or of an electrolytic cell) connected to the positive pole of a battery. Through this electrode the electrons leave the system.

Antimatter: Matter consisting of antiparticles. For example an antihydrogen consists of one positron (antielectron) and one antiproton.

Antiparticle: A particle with the same mass and spin as another particle, but with equal and opposite electric charge (and other quantum numbers are opposite which are not considered in this book). For example, the antiproton is the antiparticle of the proton while the positron is the antiparticle of the electron. When particle and antiparticle meet, they annihilate each other producing photons.

Astronomical unit (abbreviated by AU): A unit of length which equals the average distance from the Earth to the Sun. Its value is about 150 million kilometers.

Atom: Comes from the Greek word 'atoma' meaning something that cannot be divided. The smallest particle of an element that keeps its identification, namely, its chemical properties. The atom is composed of electrons and of a nucleus which contains most of the atomic mass. The atoms are electrically neutral. The electrons move around the nucleus in orbits, as the planets go round the Sun. The atom is the heart of all matter.

Atomic bomb: A violent explosive where energy is liberated from the nuclear fission process of heavy atomic nuclei (such as uranium).

Atomic energy: Energy derived by converting mass into energy according to Einstein's equation. This term is also used to describe the energy of an atom including the binding energy of the electrons in an atom (atomic binding energy) and the binding energy of the nucleons in the nuclei (nuclear binding energy).

Atomic number: The number of protons in an atom. This number also equals the number of electrons in a neutral atom.

Atomic weight: The average weight of the different natural isotopes of any specific element.

Aurora: Electrical discharges and plasma formation at the upper atmosphere above the Earth's North and South Poles. Charged particles of plasma arriving from the solar wind excite the air molecules causing the emission of light.

Battery: A single cell (or group of connected cells) that furnishes an electric current when introduced in a closed circuit.

Beta: The ratio of the plasma thermal pressure to the magnetic pressure (exerted by magnetic fields which confine the plasma) inside the plasma.

Beta particle (or beta ray): A particle which can be emitted by a radioactive atomic nucleus. This particle is an electron, which does not exist inside the nucleus but is created through radioactive emission.

'Big Bang': The theory explaining the genesis of our Universe.

Break-even: The term used in a thermonuclear fusion device where the released fusion energy equals the energy invested to create the hot plasma.

Bremsstrahlung (meaning braking radiation in German): Electromagnetic radiation emitted by charged particles (such as electrons) which change their velocity due to collisions with neighboring particles.

Carbon cycle: A sequence of thermonuclear fusion reactions and spontaneous radioactive decays which convert matter into energy in the form of radiation and high-speed positrons and neutrinos. (The carbon plays the role of a catalyzer.) These processes are regarded as one of the principal sources of energy of the Sun and other similar stars.

Cathode: The electrode of a vacuum tube (or an electrolytic cell) connected to the negative pole of a battery. The electrons enter the system through this electrode.

Cathode rays: The rays moving from the cathode towards the anode in a vacuum tube. These rays consist of electrons. (In physics there are also anode rays. These rays were not dealt with in this book.)

Celsius or Centigrade (temperature scale abbreviated by the letter C): A temperature scale where the freezing point of water is 0 degrees and the boiling point is 100 degrees.

Center of mass: A definite point that represents the average position of all the particles in a system. For example, for a plasma confined inside a vessel of a moving train, the center of mass is somewhere inside the vessel, independently of the train's motion.

Cepheids: Bright stars which vary their luminosity in a periodic way. These stars are used to estimate the distances of relatively near galaxies.

Chain reaction: Neutrons produced in the nuclear fission reaction cause the fission of other nuclei releasing more neutrons, and thus contribute to the propagation of this process.

Charge (electric): A property possessed by some particles causing them to exert a force on one another. The elementary unit of charge is possessed by the electron, which was arbitrarily decided to have a negative sign. Similar charges repel one another while opposite charges attract one another (see Coulomb force).

Classical physics: The subject of physics excluding relativistic and quantum effects.

Coherent waves: Waves with equal frequencies whose phases are determined in relation to each other (at a given place and at a given time).

Collective motion: The cooperative motion of particles rather than the individual motion of a particle.

Collision frequency: The average number of collisions of a given particle (any particle that one chooses to follow) with the other particles of the medium in a unit of time (say in one second).

Compression: Squeezing and increasing the density of a state.

Confinement time: The time that a plasma is confined by a magnetic field before it diffuses and hits the wall of the vessel in magnetic confinement fusion. When small pellets are compressed by lasers or ion beams nuclear fusion occurs before the pellet explodes. The duration of the nuclear fusion in this scheme is referred to as the inertial confinement time.

Conservation law: A law dictating that the total value of a quantity does not change. For example, the energy of an isolated system is unchanged.

Corona: The hot and dilute plasma surrounding the Sun. The corona is much hotter than the photosphere which surrounds the Sun's core.

Coulomb barrier: In order to fuse two positively charged nuclei (such as deuterium and tritium) it is necessary to overcome the Coulomb repulsion forces. This repulsion is called a Coulomb barrier. In order to overcome this barrier it is necessary to heat the plasma to many millions of degrees Celsius.

Coulomb force: The force acting between two electrical charges. This force is proportional to the size of the electric charges and decreases with increasing distance between the charges.

Critical surface: The place inside the plasma where the irradiating electromagnetic field's frequency is equal to the plasma oscillating frequency.

Cross section: An effective area which describes the probability for a particular collision process to occur.

Debye length (or Debye radius)**:** A length inside the plasma which describes the maximum distance at which a given electron will still be influenced by the Coulomb forces of a given positive ion. This distance decreases with increased plasma density while it increases for increased plasma temperature.

Debye shielding: The electrons surrounding an ion form a shield. This shield weakens the influence of the Coulomb force.

Decay: The disintegration of a radioactive substance due to nuclear emission of alpha or beta particles or gamma rays.

Density: Describes the number of particles in a unit volume. One can also define 'mass density' as the mass of the unit volume, 'charge density' as the charge of the unit volume, etc.

Deposition: A surface reaction on a material that causes selective deposition.

Deposition, chemical vapor (CVD): Deposition using chemical vapors.

Deposition, plasma-assisted chemical vapor (PACVD): Chemical vapor deposition in a plasma reactor.

Deuterium: An isotope of the hydrogen family denoted by the letter D. The nuclei of deuterium consist of one neutron and one proton.

Diffusion: The dispersion of plasma (or other matter) inside a medium. Scientists are concerned by the diffusive leakage of hot plasma towards the wall in thermonuclear experiments.

Dilute gas: A gas significantly less dense than the Earth's atmosphere and used in vacuum tubes for electrical discharges and for magnetic confinement fusion.

Discharge, arc: A high current and a low voltage between anode and cathode create an arc discharge inside a vacuum tube.

Discharge, corona: Very low currents and very high voltages between anode and cathode create a corona discharge inside a vacuum tube.

Discharge, glow: Low currents and high voltage between anode and cathode create a glow discharge inside a vacuum tube.

Discharge, microwave: The same as RF discharges but with a higher frequency of electromagnetic fields (a few centimeters wavelength).

Discharge, radio frequency: A radio frequency power creates a plasma inside a vacuum tube either inductively or capacitively.

Display panels, plasma: In a plasma display panel the light of each picture element (pixel) is emitted from a plasma created by an electric glow discharge.

Dissociation: The breakdown of a molecule into its constituent atoms.

Distribution (of energy): The number of particles (in a plasma, the number of electrons or ions) for every specific energy. For higher temperatures the distribution becomes more scattered. One also describes distributions of particle velocity, particle density, etc.

Doppler effect: The change in frequency of a wave due to the motion of the source which creates the wave (for example, a train whistle) or of the receiver of the wave.

Double layer: Occasionally, in a plasma two layers of opposite charge can be formed. Very strong electric fields can exist between these two layers that strongly influence the motion of the electrons in the plasma.

Electric current: The flow of electrical charges (electricity) analogous to the flow of a stream of water.

Electric field: Any charge affects the motion of all other charges by an electric force. This force is described schematically by lines of force throughout space.

Electrical discharge: The passage of an electric current through a gas. This phenomenon is usually associated with plasma creation during the electrical discharge.

Electrical sheath: Inside a plasma discharge close to the cathode or the anode, a strong change is created in the voltage, inducing locally a strong electric field.

Electricity: A fundamental entity of nature consisting of negative and positive charges. The elementary unit of the (negative) charge is that of an electron. The science of electricity deals with the laws describing the phenomena of charged particles.

Electrolytes: Solutions in which electricity can flow.

Electromagnet: When an electric current flows through coils (or conducting bars) it produces a magnetic field.

Electromagnetic force: The attraction and repulsion forces between particles of stationary or moving electric charges. The forces that keep a solid or a liquid together. The electromagnetic force has an infinite range of interaction and its strength between charged particles is extremely more powerful than that of the gravitation force.

Electromagnetic radiation: Waves of electric and magnetic fields.

Electromagnetic spectrum: The complete set of wavelengths of the electromagnetic radiation containing gamma rays, X-rays, ultraviolet, visible and infrared radiation, microwaves and radio waves.

Electron: The lightest massive elementary particle which has a negative charge. The electrons are the constituents of all the atoms and of plasma. The electrons are the carriers of electricity in most substances. All chemical properties of atoms and molecules are determined by their electrons.

Electron, bound: The electrons that move around the nucleus in an atom or in a molecule.

Electron, free: The electrons that move inside a plasma, or the electrons that move inside a solid in order to create the electric current in metals.

Electronic shell: The electrons that move around the nucleus of an atom are arranged in 'shells'. The inner shells contain the electrons near the nucleus while the outer shells describe the electrons which are far from the nucleus.

Electrostatic: Referring to the laws between charged particles or charged objects which are not in motion.

Element: A substance which cannot be decomposed by chemical means into simpler substances. The atoms described by Mendeleyev's Periodic Table are elements.

Energy: The capability of doing work. The energy content has many forms. For example, energy of motion, energy due to the position of a body (potential energy), electric energy, heat energy, mass energy, etc, can be converted from one form to another. Energy can neither be created nor destroyed and therefore the total amount of energy in an isolated system remains constant. This law is known as the 'conservation of energy'.

Equilibrium plasma: Also known as thermal plasma. In an equilibrium plasma the temperature of the electrons equals the temperature of the ions.

Etching: A surface reaction on a material that causes selective evaporation.

Etching, dry: Etching obtained with plasmas.

Etching, wet: Etching obtained with liquid chemicals such as acids.

Excited state: The state of an atom or a molecule, which has a higher energy level than its ground state (equilibrium state).

Field: Influence of a force. It can be represented pictorially by a set of lines of force. The density of these lines at any given point in space represents the strength of the field at that point. The direction of the lines equals the direction of the force associated with the field.

Fission: The splitting of an atomic nucleus into two fragments of almost equal size. For example, the uranium nucleus can be split by a neutron into two fragments.

Fluid mechanics: The science which describes gases and liquids at rest and in motion. Plasma is also described by the laws of fluid mechanics.

Force: The force acting on a body changes its velocity. Newton's laws of motion describe the force as the rate of change of momentum (mass times velocity). Equivalently, the force is proportional to the product of the mass of the body and its acceleration.

Free radicals: Highly reactive chemical atoms or molecules.

Frequency: The number of wavelengths that pass through a definite point in a unit of time (for example, during a second). High-frequency waves have short wavelengths.

Fusion: The fusing process of two light nuclei into a heavier nucleus resulting in a release of energy.

Gamma rays: Electromagnetic waves (i.e. photons) emitted from radioactive nuclei. The energy of these photons is larger than the energy of X-ray photons.

Gas: A state of matter in which the molecules move 'freely' without being restricted to some position in space due to their neighbors' influence. A gas does not have a definite shape and inside a vessel it takes the shape of the vessel. The third state of matter.

Gravitation force: The forces of attraction between masses, quantitatively summed up by Newton's law of gravitation. This law is universal.

Gun, plasma: A powder is heated and melted in a plasma discharge and is accelerated by a plasma jet toward a substrate. This can be done in air, vacuum and even under water.

Gyration radius (also called Larmor radius): A charged particle moves in a magnetic field in a circular motion. The radius of this circular motion increases for larger masses and decreases for larger magnetic fields.

Half life: The average time required for half the amount of radioactive material to decay.

Heat: The energy associated with the random motions of molecules, atoms, electrons or ions.

Helium: A light, colorless, inert, gaseous element. Its atom consists of two electrons and a nucleus with two protons and one or two neutrons. (The main isotopes are ^{4}He and ^{3}He.) The Sun, the stars and our Universe consist of about 25% helium.

Hologram: The recording of a picture in three dimensions, usually by means of lasers.

Hubble constant: According to the theory of the expanding Universe (or equivalently the Big Bang theory) the external galaxies are receding from each other and from our own Galaxy. Hubble's constant is the ratio between the recessional velocity of a distant galaxy and its distance from Earth.

Hydrogen atom: The lightest of all atoms. It is composed of one proton and one electron. The Sun, the stars and our Universe consist of about 75% hydrogen.

Hydrogen bomb (fusion bomb): A fission bomb (known as an atomic bomb) is surrounded by fusion material, such as lithium or deuterium. The fission energy is used to ignite the nuclear fusion reactions of the hydrogen isotopes. During this process enormous quantities of energy are released violently.

Ignition: A condition where the thermonuclear fusion energy equals the energy losses in the hot plasma through radiation.

Implantation: Reactions on a material resulting in selective deposition on or just below the surface.

Implantation, ion: A plasma source can be used to extract ions in order to implant them into the materials to be modified.

Inert gases: Gases that have a very low rate of chemical reactions.

Inertial confinement fusion: A method of producing nuclear fusion energy by compressing a small pellet to very high densities. The compression is usually accomplished by very intense laser or ion beams.

Infrared radiation: Electromagnetic waves with wavelengths close to, but larger than, visible light and smaller than those of microwaves.

Instability: An unstable motion. For example, plasma waves can become unstable and disintegrate.

Integrated circuits (chips): One tiny device capable of many electronic functions.

Ion: An atom that has lost or gained one or more electrons. It comes from the Greek word meaning wanderer.

Ionization: The process causing atoms to lose or gain electrons. The ionization process is usually caused by heating matter to high temperatures or through electrical discharges.

Ionosphere: The atmospheric region where a significant portion of atoms and molecules are ionized, forming a plasma which affects the propagation of radio waves.

Isotopes: Two or more nuclei containing an identical number of protons (same atomic number) but different numbers of neutrons. About 70 out of the 92 natural elements on Earth are present as mixtures of two, three or more isotopes. Tin, for instance, has as many as ten isotopes.

JET: Joint European Torus, a large Tokamak device built in England where a European collaboration is based in order to produce fusion energy by using deuterium–tritium fuel.

Kelvin: The temperature scale used in science. In this temperature scale the absolute zero is equal to −273.15 degrees Celsius. The quantity of one degree Kelvin is equal to one degree Celsius.

Kepler's laws: The laws describing the motion of the planets around the Sun.

Laser: Acronym for 'light amplification by stimulated emission of radiation'. Also an optical device that amplifies or generates coherent electromagnetic waves. Laser light can be produced at high intensity and it can be focused on small areas.

Lawson criterion: The requirement that the product of plasma density and its confinement time must exceed a particular value after the plasma has been heated to an appropriate temperature so that the energy from fusion reactions equals the energy invested in heating the plasma.

Limiter: Heavy bars of graphite (or other appropriate material) protruding from the magnetic confinement vessel wall built for preventing impurities that cause energy losses.

Line of force: An imaginary line describing an electric or magnetic field. Its direction at every point is proportional to the direction of the force.

Liquid: The second state of matter. Matter that flows freely like water and takes the shape of its container.

Lithium: The third element in Mendeleyev's Periodic Table (containing three protons). It is a source for tritium production. When lithium is irradiated by neutrons it undergoes a nuclear reaction where tritium is created, which is necessary for the thermonuclear fusion reactions.

Low temperature plasma: Low temperature plasma in nuclear fusion, for example one million degrees, is many orders of magnitude higher than high temperatures used in industry.

Macro: Large.

Macroinstabilities: Large-scale instabilities which can lead to a complete loss of plasma confinement.

Magnet: A material possessing the property of attracting iron. A magnet is also capable of affecting the motion of a charged particle.

Magnetic bottle: A vessel confining a plasma by an arrangement of magnetic fields.

Magnetic confinement fusion: A method of producing nuclear fusion energy by confining a very hot plasma (100 million degrees Celsius or more) by magnetic fields (in a magnetic bottle).

Magnetic field: An electric current or a permanent magnet affects the motion of moving charged particles through a magnetic field. The magnetic forces are described schematically by lines of force throughout space.

Magnetic mirror: A configuration of magnetic fields which act like a mirror on moving charged particles.

Magnetosphere: The region enclosing the Earth's magnetic field. The Earth's magnetosphere results from the interaction of its magnetic field and the solar wind. Most of the planets and stars have their own magnetosphere.

Mass: This is the amount of material in an object: the inertial resistance of a body to acceleration. On the Earth's surface it is equal to the weight, that is the force of gravity acting on the body. The mass is a property of an object which is not changed by a force acting on it. For example, in space, the mass is the same as here on Earth but the object is weightless.

Mass number: The total number of protons and neutrons in an atom.

Matter: Everything that man can survey, touch or feel. Whatever occupies space. Even man himself is matter.

Mean free path: The average distance traversed by a given particle (i.e. a particle chosen at random) between collisions with other particles in the medium in which it moves.

Micro: Small.

Microinstabilities: Small-scale (relative to the size of the plasma system) instabilities which usually increase the drifts of electrons and the diffusion of the plasma. These instabilities can occasionally cause a chaotic motion in the plasma, resulting in a loss of confinement.

Microwaves: Electromagnetic wavelengths larger than those of infrared and smaller than radio waves.

Molecule: Two or more atoms make up a molecule. For most substances this is the smallest unit of matter which can exist separately and retain its chemical properties.

Momentum: A quantity describing the motion of a particle (or of a body) measured by the product of its mass and velocity. The rate of change with time of the momentum is equal to the force acting upon the particle (or the body). This is Newton's famous second law.

Monomers: Units of one compound.

Neutral beams: Intense beams of neutral atoms which have large energies. These beams are used to heat the plasma in magnetic confinement devices in addition to other heating sources.

Neutrino: An electrically neutral particle of zero (or very small) mass which possesses a spin similar to the electron. The neutrino or its antiparticle, the antineutrino, is created in beta decay processes. Neutrinos interact with matter only through weak or gravitational interactions, thus enabling them to penetrate matter very easily without being affected.

Neutron: An elementary particle having no electric charge and a mass slightly larger than that of the proton. The neutrons together with the protons are the constituents of the nucleus. A free neutron is an unstable particle and it decays with a mean lifetime of 151 minutes. However, inside the nucleus the neutron is stable.

Neutron star: Extremely dense star composed of neutrons. They are believed to be created from a supernova explosion.

Non-equilibrium plasma: A plasma state where the electron temperature is not equal to the ion temperature.

Nuclear fission reactor: An electric plant whereby energy is obtained through a chain reaction of fissionable material (such as uranium) in a controlled way. Significant electric power is supplied by nuclear reactors in European countries, in the USA, Canada and Japan (from a few per cent to a few tens of per cent, depending on the country). There are approximately 500 nuclear fission reactors in operation today.

Nuclear fusion: see fusion.

Nuclear fusion reactor: An electric plant where energy will be obtained from controlled thermonuclear fusion reactions. Nuclear fusion reactors are not yet in operation.

Nucleon: The name for a neutron or a proton inside the nucleus.

Nucleus (plural **nuclei**)**:** The core of the atom containing most of its mass. The nucleus consists of protons and neutrons.

Ohm: The unit used to measure the resistance to an electric current.

Ohm's law: The electrical current in a conductor is proportional to the potential difference between its ends. The ratio between the potential and the current is the resistance.

Ohmic heating: The heating of a system by the passage of an electrical current. Due to the resistance of the system, the energy associated with the current flow is transformed into heat. This is the main heating source in a Tokamak.

Oscillation: A vibration. The electrons and ions oscillate independently in a plasma.

Pellet: A small fusion fuel target usually contained within multiple concentric spheres of glass, plastic, gold and other materials. The pellet is irradiated by lasers or ion beams in order to compress the fusion fuel to high densities to achieve thermonuclear fusion reactions.

Periodic Table: The classification of chemical elements in the order of their atomic numbers. The elements show a periodicity of properties with chemically similar elements recurring in a definite order, known also as the Mendeleyev table, after its inventor.

Photon: A massless uncharged particle associated with an electromagnetic wave which has energy and momentum and always moves with the speed of light.

Photosphere: The outer layer surrounding the Sun's core which can be seen from the Earth.

Pinch: A plasma column carrying a large electrical current tends to contract in the radial direction; the plasma is 'pinched' or 'squeezed' due to the interaction between the current in the plasma and the magnetic field that this current creates.

Plasma: Charged particles, namely electrons and ions, governed by electric and magnetic forces which possess collective behavior. Most of our Universe is composed of plasma, including our Sun. Plasma is described as the fourth state of matter.

Plasma polymerization (PP): The deposition of organic materials (polymers), such as unsaturated fluorocarbon gas, in a plasma reactor.

Plasma reactors: A vacuum vessel containing plasma for the use in industry.

Polymers: Many repeated units of one or more compounds.

Population inversion: This is when there are more electrons in the upper state of an atom or molecule than in the lower state.

Positron: The antiparticle of the electron. It has the same mass as the electron but possesses a positive charge equal in magnitude to the electron charge. When a positron collides with an electron, the positron–electron pair is annihilated and two or three photons are created.

Potential: A system containing electric charges or currents is also described by a field that has an absolute value lacking any direction (unlike the electric and magnetic fields). It is defined as the energy invested in order to carry a positive charge of one unit from infinity to the specific point being measured.

Power: The quantity of work carried out in a unit of time (one second).

Pressure: The force applied per unit area. Unlike the force, the pressure has an absolute value, lacking any direction. Plasma pressure is proportional to its density times its temperature.

Proton: The positively charged particles which are the constituents of the nucleus (together with the neutrons). The proton is about 1836 times heavier than the electron.

Pulsar: A class of astronomical objects believed to be neutron stars of small dimensions. The material is extremely compressed so that the electrons combine with the protons to form neutrons. Pulsars spin very rapidly (about one turn per second or even faster) and emit energy in the form of radio waves. Pulsars are believed to be formed during the explosion of a supernova.

Quantum: Discrete natural unit of energy, of electric charge or of some other physical property. For example, light is described by photons which are quanta of energy. All phenomena in the realm of quantum mechanics (submicroscopic systems) exhibit the phenomena of quantization.

Quantum mechanics: The mathematical physics that describes the motion of electrons, protons, neutrons and atomic particles. The discipline is used both in chemistry and physics to describe atoms, molecules and the solid state of matter.

Quasineutral: Almost neutral. A plasma where the charges of the electrons almost equal the charges of the ions.

Radiation: Any form of energy propagated as rays, waves or a stream of particles.

Radioactivity: The spontaneous disintegration of the nuclei of some of the isotopes. During the disintegration the elements can emit alpha or beta particles and gamma rays.

Relativity (general theory of): In 1915 Einstein published the generalization of the special theory of relativity. In 1905 he described systems moving with constant velocity relative to each other. In 1915 he described also the motion of accelerating bodies. In this theory Einstein assumes the equivalence between acceleration and gravitational forces.

Relativity (special theory of): A theory developed by Albert Einstein in 1905. Einstein assumes (a) that the laws of nature are the same for systems moving with constant velocity relative to each other, and (b) that the velocity of light is the same constant in all such systems. Consequences from these assumptions are: (1) the contraction of moving objects; (2) time extension in a moving frame; (3) increase in mass due to its velocity; and (4) the equivalence between mass and energy.

Resistance: Measured in units of ohms. In a plasma the resistance is proportional to the collision frequency.

RF (radio frequency) waves: Electromagnetic waves used for radio transmissions (between 1 and 100 MHz) and plasma devices.

Semiconductor: A material, such as germanium or silicon, which is at room temperature not as good an electrical conductor as the metals and not as bad as an insulator.

Semiconductor electronics industry: The production of almost all of the high technology electronic devices such as computers, televisions, radio, cellular phones, etc.

Solar wind: Streams of plasma matter, usually electrons and protons, flowing from the Sun outwards. The solar wind affects the Earth's magnetosphere and in particular the auroral phenomena and the Van Allen radiation belts.

Solid: A state of matter in which the motions of the molecules (or atoms) are restricted by the forces acting upon them due to their neighbors. The molecules (or atoms) are retained in a fixed position relative to each other so that usually the solid has a crystal structure. A solid has a definite shape and volume. Often referred to as the first state of matter.

Space propulsion, plasma: The principle is based on ejecting electrical heated plasmas out of the vehicle thus requiring less fuel mass than in existing chemical systems. Since the higher the temperature, the faster the speed, increasing the velocity of the plasma jet decreases the weight of the fuel.

Spectrometer: A device measuring the emitted spectrum from an object such as a star, a plasma, etc.

Spectrum: Any particular distribution of electromagnetic radiation, such as the colors produced when sunlight is dispersed through a prism. Every element, molecule or substance has a particular distinctive spectrum.

Spin: Certain elementary particles (such as the electron, the proton and the neutron) are characterized by a quantity which describes their rotation about an axis (similar to the Earth spinning about its north–south axis).

Spray, plasma: Used to apply a thick coating to substances that do not yield to other treatments. In this way an effective anti-corrosion coating is achieved.

Star: A self-luminous celestial body, such as our Sun. The source of energy of the stars is thermonuclear fusion reactions.

Statistical mechanics: The physics that describes the laws of motion of a system containing many particles through average estimations. The laws of statistical mechanics are also used in plasma physics.

Strong force: This force between the constituents of the nucleus (protons and neutrons) is responsible for the nuclear structure of the atoms. This force is about a hundred times stronger than the electromagnetic force; however its range is extremely short, about the dimensions of the nucleus only (which is a hundred thousand times smaller than atomic dimensions). Outside the nucleus the force is zero.

Superconductors: Electric conductors that have no resistance to an electric current and carry powerful currents without losses. Thus these conductors can produce strong magnetic fields without requiring a large supply of electric power. Usually these properties occur at very low temperatures. In recent developments in this research materials have been found that become superconductors at higher temperatures than previously known. Moreover, there is now hope for the development of new superconductors that work at room temperature.

Temperature: A degree of hotness or coldness measured on a definite scale. The most frequently used temperature scale is the Celsius (or centigrade) scale; one degree Celsius is 1/100 of the difference between the temperatures of boiling water and melting ice. The scientific temperature is the Kelvin scale where the unit is equal to that of the Celsius scale but the absolute zero is equal to −273.15 degrees of the Celsius scale. In statistical mechanics the temperature is measured as the spread of the energy distribution.

Thermonuclear fusion: Nuclear fusion reactions between the light elements which are caused by burning the fusion fuel (e.g. deuterium, tritium, etc).

Tokamak: A torus (doughnut, automobile tire) shaped magnetic bottle considered today to be the best performing magnetic fusion device.

Torch, plasma: See gun, plasma.

Torus: A magnetic device in the shape of a doughnut.

Tritium: The unstable heavy isotope of hydrogen denoted by the letter T. The nucleus of tritium consists of two neutrons and one proton.

Ultraviolet radiation: Electromagnetic waves with wavelengths close to and shorter than those of visible light.

Uranium: A heavy radioactive element. The last natural element in Mendeleyev's Periodic Table. It contains 92 electrons and 92 protons. The two main isotopes of uranium have a mass number of 235 and 238 where the latter is present in an abundance of 99.28%.

Vacuum tube: A sealed device containing dilute gas in which electrons move between two electrodes (two metal rods or wires) connected to a battery (or any other electric outlet).

Van Allen belt (or Van Allen radiation belt): The doughnut-shaped zone beginning approximately at altitudes of 1000 kilometers where energetic charged particles (such as electrons and protons) surround the Earth.

Volt: The practical unit of measuring electrical potential difference.

Watt: The practical unit of power (work done per unit time).

Wave: A wave is a disturbance propagating in a medium.

Wavelength: In a wave the wavelength is the distance between two successive wave crests. In particular, for electromagnetic waves this is the distance between successive points where the electric (or magnetic) field has a maximum value. The shorter the wavelength the larger is its frequency.

Weak force: This force manifests itself through radioactive beta decay. Interactions between neutrinos are purely of this type. The range of the weak force is extremely small (of the order of nucleon dimensions) and its strength is much smaller than the electromagnetic force. About three decades ago the weak and electromagnetic forces were unified and are thus described mathematically within the same formalism.

Weight: The force with which a body is attracted towards the Earth.

Work: The product of a force and the distance along which this force is acting.

X-rays: Electromagnetic radiation of wavelengths much shorter than those of ordinary (visible) light but longer than gamma rays. This radiation is usually produced when high-energy electrons impinge upon a metal target.

X-ray laser: Lasers with very small wavelengths which enable one to 'view' individual molecules in living tissues. These can also be used for three-dimensional viewing of any small object in the order of the laser wavelength.

Z-pinch: A straight line electric current passing through the plasma forces the plasma to be squeezed (see also pinch).

Bibliography

General Public

Artsimovich, L. A. (1965) (originally published in Russian 1963) *Elementary Plasma Physics.*

'BP-Amoco Statistical Review of World Energy, June 1999' http://www.bpamoco.-com

Bradu, P. (1999) *L'Univers des Plasmas du Big Bang aux Technologies du 3e Millenaire,* Paris: Flammarion.

Bruce, J., Lee, H. and Haites, E. (eds) (1996) *Climate Changes 1995: Economic and Social Dimensions of Climate Change,* Cambridge: Cambridge University Press.

Dolan, T. J. (1982) *Fusion Research,* New York: Pergamon Press.

Eliezer, Y. and Eliezer, S. (1989) *The Fourth State of Matter: An Introduction to the Physics of Plasma,* Bristol: Institute of Physics Publishing.

Fowler, T. K. (1997) *The Fusion Quest,* Baltimore, MD: Johns Hopkins University Press.

Goodman, B. (1998) 'Star in a bottle', *Air and Space,* February/March, p. 68.

Hawking, S. W. (1998) *A Brief History of Time,* New York: Bantam Books.

Herman, R. (1991) *Fusion, The Search for Endless Energy,* Cambridge: Cambridge University Press.

Lang, K. R. (1997) *Sun, Earth and Sky,* Berlin: Springer.

Rhodes, R. (1995) *Dark Sun. The Making of the Hydrogen Bomb,* New York: Simon and Schuster.

US National Research Council (1995) *Plasma Science: From Fundamental Research to Technological Applications,* Washington, DC: National Academic Press.

Weinberg, S. W. (1977) *The First Three Minutes, a Modern View of the Origin of the Universe,* New York: Basic Books.

Wilhelmsson, H. (2000) *Fusion, A Voyage through the Plasma Universe,* Bristol: Institute of Physics Publishing.

Wynn-Williams, G. (1992) *The Fullness of Space,* Cambridge: Cambridge University Press.

Introductory

Akasofu, S. I. and L. J. Lanzerotti (1975) 'The Earth's magnetosphere', *Physics Today,* **28**, 12.

Alfven, H. (1986) 'The plasma universe', *Physics Today,* **33**(9) 22.

Brown, S. C. (1978) 'A short history of gaseous electronics', In Hirsch, M. N. and Oskam, H. J. (eds), *Gaseous Electronics*, Vol. I, New York, Academic Press, pp. 1–18.

Chapman, S. (1968) 'Historical introduction to aurora and magnetic storms', *Annales de Géophysique*, **24**, 497.

Chen, F. F. (1990) *Introduction to Plasma Physics*, New York: Plenum Press.

Dendy, R. (1993) *Plasma Physics; an Introduction Course*, Cambridge: Cambridge University Press.

Eliezer, S. (1992) 'Laser fusion for pedestrians', Israel Atomic Energy Commission Report, IA-1374.

Goldston, R. J. and Rutherford, P. H. (1995) *Introduction to Plasma Physics*, Bristol: Institute of Physics Publishing.

Harms, A. A., Schoepf, K. F., Miley, G. H. and Kingdon, D. R. (2000) *Principles of Fusion Energy*, Singapore: World Scientific.

Hazeltine, R. and Waelbroeck, F. (1998) *The Framework of Plasma Physics*, Parseus Books.

Hogan, W. J., Bangerter, R. and Kulcinski, G. L. (1992) 'Energy from inertial fusion', *Physics Today*, September, p. 42.

Houghton, J. T., Jenkins, G. J. and Ephraums, J. J. (eds) (1990) *Climate Change: The Intergovernmental Panel on Climate Change (IPCC) Scientific Assessment*, Cambridge: Cambridge University Press.

ITER Fusion Research/Steady Progress, http://www.iter.org.

Kapitza, P. L. (1979) 'Plasma and controlled thermonuclear reaction', *Science*, **205**, 959.

Kaw, P. K. (1993) 'Fusion power: who needs it?' in *Plasma Physics and Controlled Nuclear Fusion Research, 1992*, Vol 1, Vienna: IAEA.

Key, M. (1991) 'Laser generate plasma power', *Physics World*, August, p. 52.

Lawrence Livermore National Laboratory (LLNL) Laser program, http://www.llnl.gov.

More, T. R. (1992) 'Atoms in plasmas', *Physics World*, April, p. 38.

Mourou, G. A., Barty, C. P. J. and Perry, M. D. (1998) 'Ultrahigh-intensity lasers, physics of the extreme on a tabletop', *Physics Today*, January, p. 22.

Nuckolls, J. H., Woods, L., Thyssen, A. and Zimmermann, G. (1972), 'Laser compression of matter to super-high densities: thermonuclear applications', *Nature*, **239**, 139.

Ongena, J. and Van Oost, G. (2000) *Transaction of Fusion Technology*, **37**, pp. 3–15. *International Energy Annual 1995* (1996) Report DOE/EIA-0219, Washington, DC, http://www.eia.doe.gov.

Proud, J. *et al.* (1991) *Plasma Processing of Materials: Scientific Opportunities and Technological Challenges*, Washington, DC: National Research Council, U.S. Department of Commerce N9217556.

Rose, S. (1994) 'Astrophysical plasma laboratories', *Physics World*, April, p. 56.

Rosen, M. D. (1999) 'The physics issues that determine inertial confinement fusion target gain and driver requirements: a tutorial', *Physics of Plasma*, **6**, 1690.

Visualization of MHD (magneto hydrodynamics) Phenomena in Tokamaks and Stellarators, http://www.oornl.gov.fed.mhd/mhd.html.

Yonas, G. (1998) 'Fusion and the Z-Pinch', *Scientific American*, August, p. 23.

Advanced

Alfven, H. (1950) *Cosmic Electrodynamics*, Oxford: Oxford University Press.

Alfven, H. (1981) *Cosmic Plasma*, Dordrecht: Reidel.

Alfven, H. and Falthammar, C. G. (1963) *Cosmic Electrodynamics*, Oxford: Clarendon.

Berk, H. I. (1993) 'Fusion, magnetic confinement', in *Encyclopedia of Applied Physics*, Vol. 6, pp. 575–607, VCH Publishers.

Boulos, M., Fauchais, P. and Pfender, E. (1994) *Thermal Plasma Processing*, Vol. 1, New York: Plenum Press.

Brown, S. C. (1966) *Introduction to Electrical Discharges in Gases*, New York: Dover.

Bryant, D. (1999) *Electron Acceleration in the Aurora and Beyond*, Bristol: Institute of Physics Publishing.

Cairns, R. A. (1991) *Radiofrequency Heating of Plasmas*, Bristol: Institute of Physics.

Cobine, J. D. (1958) *Gaseous Conductors*, New York: Dover.

Cross, R. (1988) *An Introduction to Alfven Waves*, Bristol: Institute of Physics Publishing.

Davidson, R. (1990) *An Introduction to the Physics of Nonneutral Plasma*, New York: Addison-Wesley.

Deutsch, C. (1986) 'Inertial confinement fusion driven by intense ion beams', *Annales de Physique*, **11**, pp. 1–111.

Eliezer, S. and Ricci, R. A. (eds) (1991) 'High pressure equations of state: theory and applications', in *Proceedings of the International School of Physics 'Enrico Fermi'*, Course 113, Amsterdam: North-Holland.

Eliezer, S., Ghatak, A. and Hora, H. (1986) *An Introduction to Equations of State: Theory and Applications*, Cambridge: Cambridge University Press.

Flemings, C. M. *et al.* (1985) *Plasma Processing of Materials*, Washington, DC: National Materials Advisory Board, U.S. Department of Commerce, ADA 152398.

Freidberg, J. P. (1987) *Ideal Magnetohydrodynamics*, New York: Plenum.

Ginzburg, V. L. (1961) *Propagation of Electromagnetic Waves in Plasma*, New York: Gordon and Breach.

Glasstone S. and Lovberg, R. H. (1960) *Controlled Thermonuclear Reactions*, New York: Van Nostrand.

Griem, H. R. (1997) *Principles of Plasma Spectroscopy*, Cambridge: Cambridge University Press.

Gross, B., Grycz, B. and Miklossy, K. (1969) *Plasma Technology*, New York: American Elsevier.

Hazeltine, R. D. and Meiss, J. D. (1992) *Plasma Confinement*, Perseus Books.

Heald, M. A. and Wharton, C. B. (1965) *Plasma Diagnostics with Microwave*, New York: Wiley.

Hess, W. N. (1968) *The Radiation Belt and Magnetosphere*, Waltham, MA: Blaisdell.

Hitchon, W. N. G. (1999) *Plasma Processes for Semiconductor Fabrication*, Cambridge: Cambridge University Press.

Hogan, W. J., Coutant, J., Nakai, S., Rozanov, V. B. and Velarde, G. (eds) (1995) *Energy from Inertial Fusion*, Vienna, Austria: International Atomic Energy Agency Publication.

Hora, H. (1981) *Physics of Laser Driven Plasma*, New York: Wiley.

Hora, H. (2000). *Laser Plasma Physics, Forces and the Nonlinearity Principle,* Bellingham, WA: SPIE Press.

Howatson, A. M. (1976) *An Introduction to Gas Discharges,* Oxford: Pergamon Press.

Hutchinson, I. H. (1994) *Principles of Plasma Diagnostics,* Cambridge: Cambridge University Press.

Ichimaru, S. (1994) *Statistical Plasma Physics,* Vol. 1, *Basic Principles,* Vol. 2, *Condensed Plasmas,* Perseus Books.

Kadomtsev, B. B. (1992) *Tokamak Plasma, a Complex Physical System,* Bristol: Institute of Physics Publishing.

Kelley, M. C. (1989) *The Earth's Ionosphere,* New York: Academic Press.

Kippenhahn, R. and Weigert, A. (1990) *Stellar Structure and Evolution,* New York: Springer.

Kirk, J. G., Melrose, D. B. and Priest, E. R. (1994) *Plasma Astrophysics,* Berlin: Springer.

Krall, N. and Trivelpiece, A. (1973) *Principles of Plasma Physics,* New York: McGraw-Hill.

Kruer, W. L. (1988) *The Physics of Laser Plasma Interactions,* Redwood City, CA: Addison-Wesley.

Liberman, M. A., DeGroot, J. S., Toor, A. and Spielman, R. B. (1999) *Physics of High Density Z-Pinch Plasmas,* New York: Springer.

Lieberman, M. and Lichtenberg, A. (1994) *Principle of Plasma Discharges and Materials Processing,* New York: Wiley.

Lindl, J. (1998) *Inertial Confinement Fusion,* Berlin: Springer.

Liu, C. S. and Tripathi, V. K. (1995) *Interaction of Electromagnetic Waves with Electron Beams and Plasmas,* Singapore: World Scientific.

MacDonald, A. D. (1966) *Microwave Breakdown in Gases,* New York, Wiley.

Manheimer, W., Sugiyama, I. and Stix, T. H. (eds) (1996) *Plasma Science and the Environment,* New York: American Institute of Physics.

Michel, F. C. (1990) *Theory of Neutron Star Magnetospheres,* Chicago: University of Chicago.

Moir, R. W., Nakashima, H., Basov, N. G., Eliezer, S., Lee, J. D. and Logan, B. G. (1995) 'Uses of inertial fusion energy technology involving implosions', in Hogan, W. J. *et al. Energy from Inertial Fusion.*

Nicholson, D. (1983) *Introduction to Plasma Theory,* New York: Wiley.

Parks, G. K. (1991) *Physics of Space Plasmas, an Introduction,* Redwood City, CA: Addison-Wesley.

Peratt, A. (1992) *Physics of the Plasma Universe,* Berlin: Springer.

Polak, L. S. and Lebedev, Yu. A. (eds) (1998) *Plasma Chemistry,* Cambridge: Cambridge University Press.

Post, R. F. (1956) 'Controlled fusion research—an application of the physics of high temperature plasma', *Review of Modern Physics,* **28**, 338.

Rosenbluth, M. N. (1994) *New Ideas in Tokamak Confinement,* Berlin: Springer.

Rosenbluth, M. N., Sagdev, R. Z., Galeev, A. A. and Sudan, R. N. (1983) *Handbook of Plasma Physics,* Vol. 1, *Basic Plasma Physics 1,* Amsterdam: North-Holland.

Rosenbluth, M. N., Sagdev, R. Z., Galeev, A. A. and Sudan, R. N. (1984) *Handbook of Plasma Physics,* Vol. 2. *Basic Plasma Physics 2,* Amsterdam: North-Holland.

Rosenbluth, M. N., Sagdeev, R. Z., Rubenchik, A. and Witkowski, S. (1991) *Handbook of Plasma Physics*, Vol. 3, *Physics of Laser Plasma*, Amsterdam: North-Holland.

Roth, J. R. (1995) *Industrial Plasma Engineering Principles*, Bristol: Institute of Physics Publishing.

Salzmann, D. (1998) *Atomic Physics in Hot Plasmas*, Oxford: Oxford University Press.

Schmidt, G. (1966) *Physics of High Temperature Plasmas, An Introduction*, New York: Academic.

Spitzer, L. (1962) *Physics of Fully Ionized Gases*, New York: Interscience.

Stacey, W. (1984) *Fusion: an Introduction to the Physics and Techniques of Magnetic Confinement Fusion*, New York: Wiley.

Stix, T. H. (1992) *Waves in Plasmas*, New York: American Institute of Physics.

Sugawara, M. (1998) *Plasma Etching: Fundamentals and Applications*, Oxford: Oxford University Press.

Tajima, T. and Shibota, K. (1997) *Plasma Astrophysics*, Reading, MA: Addison-Wesley.

Tajima, T. (1989) *Computational Plasma Physics with application to Fusion and Astrophysics*, Redwood City, CA: Addison-Wesley.

Teller, E. (ed.) (1981) *Fusion*, New York: Academic.

Velarde, G., Ronen, Y. and Martinez-Val, J. M. (1992) *Nuclear Fusion by Inertial Confinement, a Comprehensive Treatise*, Boca Raton, FL: CRC Press.

Yamanka, C. (1991) *Introduction to Laser Fusion*, London: Harwood.

Index